Looking for Lies

An Exploratory Analysis for Automated Detection
of Deception

MAREK N. POSARD, CHRISTIAN JOHNSON, JULIA L. MELIN,
EMILY ELLINGER, HILARY REININGER

Prepared for the Performance Accountability Council Program
Management Office
Approved for public release; distribution unlimited

NATIONAL DEFENSE RESEARCH INSTITUTE

For more information on this publication, visit **www.rand.org/t/RRA873-1**.

About RAND

The RAND Corporation is a research organization that develops solutions to public policy challenges to help make communities throughout the world safer and more secure, healthier and more prosperous. RAND is nonprofit, nonpartisan, and committed to the public interest. To learn more about RAND, visit www.rand.org.

Research Integrity

Our mission to help improve policy and decisionmaking through research and analysis is enabled through our core values of quality and objectivity and our unwavering commitment to the highest level of integrity and ethical behavior. To help ensure our research and analysis are rigorous, objective, and nonpartisan, we subject our research publications to a robust and exacting quality-assurance process; avoid both the appearance and reality of financial and other conflicts of interest through staff training, project screening, and a policy of mandatory disclosure; and pursue transparency in our research engagements through our commitment to the open publication of our research findings and recommendations, disclosure of the source of funding of published research, and policies to ensure intellectual independence. For more information, visit www.rand.org/about/research-integrity.

RAND's publications do not necessarily reflect the opinions of its research clients and sponsors.

Published by the RAND Corporation, Santa Monica, Calif.
© 2022 RAND Corporation
RAND® is a registered trademark.

Library of Congress Cataloging-in-Publication Data is available for this publication.
ISBN: 978-1-9774-0952-2

Cover images: I mac/Getty Images and Fizkes/Getty Images

Cover composite design: Carol Ponce.

About This Report

In this report, we document research and analysis to pilot test new techniques for detecting deception during security clearance background investigations. The study was sponsored by the Performance Accountability Council Program Management Office as a part of an overall project effort to consider new interviewing methods. The RAND Corporation subcontracted with a team from the University of New Haven under the direction of Dr. Charles "Andy" Morgan to design a series of experiments. Data collection used in this research was conducted as part of the same subcontract with the University of New Haven. In this report, we describe results from the use of computational methods to analyze a portion of the data collected under this subcontract. Additional analysis using the same dataset may be published in other non-RAND outlets on a future date.

The purpose of this report is to describe the development and testing of the potential use for machine-learning methods to detect speech patterns that reflect deceit or truthfulness during simulated background investigation interviews conducted by our subcontractor, the University of New Haven. We conclude that these exploratory models, when calibrated and pilot tested, have the potential to help the government augment its existing processes for detecting deception during the security clearance process.

The research reported here was completed in April 2022 and underwent security review with the sponsor and the Defense Office of Prepublication and Security Review before public release.

RAND National Security Research Division

This research was sponsored by Performance Accountability Council's Program Management Office and conducted within the Forces and Resources Center of the RAND National Security Research Division (NSRD), which operates the National Defense Research Institute (NDRI), a federally funded research and development center sponsored the Office of the Secretary of Defense, the Joint Staff, the Unified Combatant Commands, the Navy, the Marine Corps, the defense agencies, and the defense intelligence enterprise.

For more information on the RAND Forces and Resources Center, see www.rand.org/nsrd/frp.html or contact the director (contact information is provided on the webpage).

Acknowledgments

We thank our sponsors from the Performance Accountability Council Program Management Office, particularly for the insightful feedback that we received from David Colangelo, Renee Oberlin, Travis Furman, and Joseph Plummer. We are grateful to Melissa Bauman, whose dedicated work greatly improved the prose of this report. We are also thankful to Priya Gandhi, Nathan Thompson, and Cory Stern for their help with reviewing audio transcripts. We are grateful to Sina Beaghley, William Marcellino, and Pete Schirmer from the RAND Corporation and Maria Hartwig from the City University of New York for their considered and thoughtful reviews. We recognize our subcontractor, the University of New Haven, under the direction of Dr. Morgan, who informed the design of a series of experiments to pilot test new interviewing techniques and who also collected the data for RAND. Finally, we thank the late Beau Shanahan, Ryan Lebel, and other Research Assistants from the University of New Haven for their work on this project.

Contents

About This Report... iii
Figures and Tables.. vii
Summary.. ix

CHAPTER ONE

Introduction 1
  Study Background.. 1
  Report Outline 2

CHAPTER TWO

Relevant Background Literature 3
  Select Literature on Detecting Deception 3
  Machine Learning for Detecting Deceptive Speech............................ 8
  Summary.. 10

CHAPTER THREE

Description of Data... 13
  Sample and Procedures... 13
  Interview Technique... 15
  Analysis of Interview Data.. 17

CHAPTER FOUR

Results from Analysis of Interview Data 19
  Background of the Linguistic Models....................................... 19
  Analysis of Virtual Teleconference Interviews............................. 21
  Analysis of Virtual Chat Interviews....................................... 32
  Summary of Key Findings 38

CHAPTER FIVE

Potential Sources of Bias 39
  Research Finds Gender Might Affect How People Are Deceptive 39
  Does the Interviewer's Gender Matter?..................................... 43
  Implications and Key Considerations....................................... 44

CHAPTER SIX

Limitations, Conclusions, and Recommendations................................ 47
  Limitations... 47
  Conclusions and Recommendations 49
  Potential for Next Steps 50

APPENDIXES

A. Modified Cognitive Interviewing 53
B. Study Materials 55
C. Example Output from Amazon Web Services Transcribe 61
D. Proof of Concept: Deep Learning Contradiction Model 65

Abbreviations 75
References 77

Figures and Tables

Figures

3.1. Order of Operations for This Study 14
4.1. Feature Importance, Model 1 22
4.2. Average Frequency of Select Word Uses by Study Condition 23
4.3. Comparison of Select Stance Vectors by Study Condition 26
4.4. Word Lengths Used by People in Truthful and Deceitful Conditions 27
4.5. Words per Second and Fraction of Smaller Words for Truthful and Deceitful Speech 28
4.6. Number of Total Words and Unique Words Used by People in Liar and Truth-Telling Conditions 29
4.7. Number of Total Words and Unique Words Used by People in Liar and Truth-Telling Conditions, Chat Logs 33
4.8. Word Importance for Combined Chat and VTC Model 34
4.9. Average Frequency of Select Words Used by Study Condition, Chat Logs ... 35
4.10. Comparison of Select Stance Vectors by Study Condition, Chat Logs 37
5.1. Word Usage by Gender 42
5.2. Word Importance by Gender 43
B.1. Demographic Questionnaire for Study Participants 55
B.2. Fictitious News Story About the Leaking of Classified Information 56
B.3. Fictitious Sensitive Memorandum About the Leaking of Classified Information 57
B.4. Introductory Video Before Interview by Former FBI Special Agent 58
C.1. Screenshot of Plot from AWS Transcribe 62
D.1. Contradiction and Entailment Matrix for Respondent #363 (Liar) 67
D.2. Attention Scores for Statements for Respondent #363 (Liar) 68
D.3. Contradiction and Entailment Matrix for Respondent #351 (Truth-Teller) ... 69
D.4. Attention Scores for Statements for Respondent #351 (Truth-Teller) 70
D.5. Contradiction and Entailment Matrix for Respondent #262 (Liar) 71
D.6. Attention Scores for Statements for Respondent #262 (Liar) 72

Tables

3.1. Sample Demographics 15
4.1. Top Over-Present Stance Vectors by Study Condition 25
4.2. Classification Performance of Exploratory Linguistic Models 30
4.3. Representative Confusion Matrix for Model 1 31
4.4. Classification Performance of Models on Chat Logs 32
4.5. Top Over-Present Stance Vectors by Study Condition, Chat Logs 36
5.1. Classification Performance by Gender 41
C.1. Transcription Using AWS Transcribe 61
C.2. Partial Transcript of Interview with Participant 63

Summary

The processes for applying for, investigating, and adjudicating a security clearance in the United States are costly in terms of time and effort for both applicants and the federal government. Interviews with applicants about their backgrounds are a key component of this process. The Performance Accountability Council asked the RAND Corporation's National Security Research Division to develop and pilot test the potential for new interviewing methods, including machine learning (ML), to detect speech patterns that reflect deceit or truthfulness during simulated background investigation interviews. We propose that *how* people answer questions might be just as important as *what* they say during these interviews. Using this proposition, we developed and tested ML models to detect truthfulness and deceit among interviewees participating in a simulated background investigation interview.

How the Study Worked

Participants in the study were asked to read a story about someone leaking classified information. They were randomly assigned to read the same story, presented as either a news report or a memo with markings indicating this document contained "sensitive" information. Then participants were randomly assigned to one of two study conditions where research staff told participants to be (1) completely honest during the interview about what they had just read, or (2) deceitful about the content of the document. Participants then spoke with interviewers, who are retired law enforcement officers who are trained in using one particular interview technique—modified cognitive interviewing (MCI) (described in detail in Appendix A)—via video teleconference (VTC) and text-based chat. Participants answered the same interview questions in random order via an asynchronous interview chat (i.e., an online process with no human interviewer present) either before or after their VTC interview. Our subcontractor audio recorded these interviews, and we transcribed them for analysis. We also analyzed the text logs of the chats.

The authors developed several ML models that used features of interviewees' responses to detect who was being deceptive or truthful during this study. These models used word count, linguistic stance (e.g., the presence of characteristics for emotion or doubt), metadata (e.g., average word length), and a combination of these factors to predict deception. Results show that models that used word count produced high accuracy rates for the VTC and chat interviews and also high accuracy rates came from a combination of chat and interviews. The authors also tested a deep learning model that compares all statements made by an individual interviewee to detect potential contradictions (see Appendix D). Results from this model suggest that it could be useful in detecting attempts at deception after calibration using relevant training data. Finally, the research team analyzed differences that the model found between men and women in our sample. Results show noticeable differences in accuracy rates by gender.

The Results Show Promise for Background Checks

This report concludes that ML models are promising tools that have the capacity to augment existing security clearance background investigation processes. We outline six recommendations for the federal government: (1) testing ML modeling of interview data, (2) testing alternatives to in-person security clearance interviews (e.g., VTC and chat-based modes), (3) testing the use of asynchronous interviews via chat to augment existing interview techniques, (4) using ML tools to augment existing investigation process, (5) validating any ML models used for security clearance investigations to limit sources of bias, and (6) using a human-in-the-loop to continuously calibrate any ML models.

We conclude that these exploratory models, when calibrated and tested, could help the government augment its existing processes for detecting deception during the security clearance process.

CHAPTER ONE

Introduction

In the United States, a security clearance is a privilege, not a right. The process for determining who is eligible for this privilege is costly. Applicants must complete a 136-page questionnaire that asks various personal questions about their lives (e.g., where they have lived in the previous ten years, marital and relationship history, personal debts, history of drug use).[1] The U.S. government, in turn, spends time and resources to confirm this information and interview applicants about their past and present associates. This report presents results from an exploratory analysis that tests automated tools for detecting when some of these applicants attempt to deceive the government during the interview portion of this process.

Study Background

In 2019, the government approved confidential, secret, or top secret security clearances for an estimated 964,138 people. Historically, a relatively small percentage of applicants are denied a clearance.[2] In 2017, for example, the denial rates for new clearances ranged from zero to 5.9 percent for agencies within the intelligence community.[3] The revocation rate for those holding an existing clearance was even lower, between zero and 2.3 percent. Thus, the government is tasked with identifying a relatively small number of people who should not receive the privilege of having access to the country's classified information.

We propose that automated tools could supplement current government efforts to identify when some of these applicants are more—rather than less—likely to be denied a security clearance. Specifically, these tools might help the government detect some signals of deception when officials interview clearance applicants. To test this hypothesis, the RAND Corporation subcontracted with the University of New Haven to inform the design of a series of experiments to pilot test new interviewing techniques. The University of New Haven com-

---

[1] Office of Management and Budget, "Questionnaire for National Security Positions," Standard Form 86, revised November 2016.

[2] National Counterintelligence and Security Center, *Fiscal Year 2019 Annual Report on Security Clearance Determinations*, Washington, D.C.: Office of the Director of National Intelligence, April 2020, p. 8.

[3] National Counterintelligence and Security Center, *Fiscal Year 2017 Annual Report on Security Clearance Determinations*, Washington, D.C.: Office of the Director of National Intelligence, 2017, p. 8.

pleted all data collection as part of its subcontract with RAND. In this study, participants read a document about the leaking of classified information that was stolen from the National Security Agency. We randomly assigned half of the participants to read one of two versions of this same document; it was presented either as a news story or as a memo with red markings stating that this document contained "sensitive" information.

These participants then completed two interviews about what they had just read—one conducted via video teleconference (VTC) and the other via text-based chat—that simulate a portion of a background investigation interview. For the interviewees who read a news story, we asked them to be truthful about what they had just read. For those who read the sensitive document, we asked them to withhold certain details about the document, including its supposed sensitive nature. We will provide more-granular detail about the study design in Chapter Three.

Typically, interviews for security clearance background investigations involve two people: interviewer and interviewee. The former asks questions, and the latter gives answers that help the government decide whether this individual has an acceptable level of risk to hold a security clearance. We propose placing importance on not only *what* interviewees say but also *how* they respond. Specifically, there might be patterns in how people respond within a controlled environment that suggests they are trying to be deceptive. In this report, we present several exploratory natural language processing (NLP) models that detect signals that are associated with these patterns.

Report Outline

In Chapter Two, we present a brief overview of relevant research on deception detection. Chapter Three outlines the methods of our study. Chapter Four presents the results from our NLP models that used data from the study's VTC interviews and virtual chats. Chapter Five presents analyses that showcase potential sources of bias, focusing on gender differences in our results. Chapter Six reviews our conclusions and presents recommendations. Appendix A provides a brief overview of the interviewing technique we used in this study, called *modified cognitive interviewing* (MCI). Appendix B presents some of the study materials used by our subcontractor. Appendix C displays a sample output from our automated transcriptions. Appendix D presents a proof of concept for applying a deep learning model to our data.

CHAPTER TWO

Relevant Background Literature

In this chapter, we summarize some of the relevant literature on deception detection. We begin by discussing select studies that are focused on detecting signals of deception in interactional settings, including specific features of deceptive speech. Next, we provide a general discussion on machine learning (ML) applications for deception detection, including some common linguistic features of deceptive speech that ML approaches have managed to detect in recent research.

Select Literature on Detecting Deception

In this section, we review some of the relevant literature that discusses patterns of speech and behavior that might present when people engage in deception. The research on detecting deception is expansive, yet mixed. Although some research has found evidence of detectable patterns in deception,[1] other research shows that there is limited evidence to indicate consistent patterns of deceit.[2] While it is beyond the scope of this report to review the entirety of this research, we provide a brief overview of research identifying the presence or absence of such patterns. We then turn our attention to relevant studies that focus on three characteristics of detecting deception in face-to-face and online interactions: sentence complexity in responses that people give when trying to deceive, filler words during deception attempts, and research on deception in online communication.

---

[1] Samantha Mann, Aldert Vrij, and Ray Bull, "Suspects, Lies, and Videotape: An Analysis of Authentic High-Stake Liars," *Law and Human Behavior*, Vol. 26, No. 3, 2002; and Sara Landström, Pär Anders Granhag, and Maria Hartwig, "Witnesses Appearing Live Versus on Video: Effects on Observers' Perception, Veracity Assessments and Memory," *Applied Cognitive Psychology*, Vol. 19, No. 7, November 2005.

[2] Maria Hartwig and Charles F. Bond Jr., "Lie Detection from Multiple Cues: A Meta-Analysis," *Applied Cognitive Psychology*, Vol. 28, No. 5, 2014; Maria Hartwig and Charles F. Bond, Jr., "Why Do Lie-Catchers Fail? A Lens Model Meta-Analysis of Human Lie Judgements," *Psychological Bulletin*, Vol. 137, No. 4, July 2011; Siegfried Ludwig Sporer and Barbara Schwandt, "Moderators of Nonverbal Indicators of Deception: A Meta-Analytic Synthesis," *Psychology, Public Policy, and Law*, Vol. 13, No. 1, 2007; and Bella M. DePaulo, James J. Lindsay, Brian E. Malone, Laura Muhlenbruch, Kelly Charlton, and Harris Cooper, "Cues to Deception," *Psychological Bulletin*, Vol. 129, No. 1, January 2003.

Signals of Deception in Current Research

There is a robust and growing literature on the topic of deception detection. Researchers in this space often focus on the ways that deception is an interactional process and how peoples' expectations about others shape interaction patterns.[3] As a result, liars are likely to be influenced by the individuals to whom they are speaking and vice versa.[4] These interactional patterns have led some researchers to believe that it might be possible to detect signals of deception not only in the words that individuals use to lie, but also in the manner that liars use such words.[5]

With respect to the context of interrogational interviews, for instance, some research has shown that liars have a tendency to speak in a higher-pitched voice.[6] Other studies have found that when being interrogated by law enforcement, guilty suspects adopt a strategy of keeping their stories simple, whereas innocent suspects opt to "tell it like it happened."[7] Other cues to deception can include liars experiencing greater arousal, more emotions like guilt or fear, and a greater expenditure of effort to control their nonverbal behaviors when compared with truth-tellers.[8] Increased *cognitive load*—defined as the number of mentally complex tasks demanded of a person—is often experienced by liars.[9] This increased mental taxation could translate into longer pauses in speech or speech-related errors, the monitoring of an interlocutor's (or interviewer's) reactions, and adjusting words or behaviors.[10]

One commonly held but erroneous belief about deceit is that people who attempt to deceive display fidgety and nervous behaviors (i.e., averted gaze, more blinking, increased body movements, increased speech disturbances).[11] However, not all of these indicators bear out in either real-world or laboratory settings. A polygraph might pick up symptoms of arousal

---

[3] Bella M. DePaulo, Deborah A. Kashy, Susan E. Kirkendol, Melissa M. Wyer, and Jennifer A. Epstein, "Lying in Everyday Life," *Journal of Personality and Social Psychology*, Vol. 70, No. 5, May 1996; and Jeffrey T. Hancock, Lauren E. Curry, Saurabh Goorha, and Michael Woodworth, "On Lying and Being Lied to: A Linguistic Analysis of Deception in Computer-Mediated Communication," *Discourse Processes*, Vol. 45, No. 1, 2008.

[4] Hancock et al., 2008.

[5] Miron Zuckerman, Bella M. DePaulo, and Robert Rosenthal, "Verbal and Nonverbal Communication of Deception," *Advances in Experimental Social* Psychology, Vol. 14, 1981.

[6] Landström, Granhag, and Hartwig, 2005.

[7] Leif A. Strömwall, Maria Hartwig, and Pär Anders Granhag, "To Act Truthfully: Nonverbal Behaviour and Strategies During a Police Interrogation," *Psychology, Crime and Law*, Vol. 12, No. 2, 2006; and Maria Hartwig, Pär Anders Granhag, and Leif A. Strömwall, "Guilty and Innocent Suspects' Strategies During Police Interrogations," *Psychology, Crime and Law*, Vol. 13, No. 2, 2007.

[8] Zuckerman, DePaulo, and Rosenthal, 1981.

[9] Mann, Vrij, and Bull, 2002; and Landström, Granhag, and Hartwig, 2005.

[10] Mann, Vrij, and Bull, 2002.

[11] Leif Strömwall and Pär Anders Granhag, "How to Detect Deception? Arresting the Beliefs of Police Officers, Prosecutors and Judges," *Psychology, Crime and Law*, Vol. 9, No. 1, 2003.

(i.e., blood pressure, pulse rate, respiration, sweating). Yet those trying to deceive might not exhibit nervous behaviors because they are simultaneously experiencing other internal processes, such as cognitive load, behavioral control, or both.[12] In actuality, these internal states could manifest in *less* blinking, *fewer* body movements, and *increased* speech pauses, all of which are opposite many nervous behaviors.[13]

Because lying is an interactional process, other factors that could influence how an individual deceives include the characteristics of the interlocutor. For instance, in the context of an interview, it is possible that an interviewer's gender (relative to the interviewee's) can affect how an interviewee responds and influence their verbal and nonverbal deceptive cues.[14] Even differences in social status between interviewer and interviewee have the potential to alter voice-related deceptive cues.[15]

On the other hand, meta-analytic studies tend to contradict the idea that there are consistent, detectable cues to deception. Such studies have converged on the finding that lies are barely evident in behavior, and that cues to deception are generally scarce and weak, at best.[16] For example, one large-scale meta-analysis that examined 158 cues of deception found that the majority of these cues were not related to deception.[17] Behaviors that did indicate a systematic link to deception were generally only weakly linked with lying.[18] Another meta-analysis that investigated nonverbal indicators of deception similarly found no evidence that nonverbal behaviors increased while lying, and that behaviors commonly deemed important indicators of deception (e.g., gaze aversion) did not vary with deception.[19] Consistent with the aforementioned studies, meta-analyses of human lie detection ability have shown that people—including both lay persons and presumed experts, such as law enforcement professionals—detect lies at an accuracy rate only marginally higher than random chance.[20]

---

[12] Mann, Vrij, and Bull, 2002.

[13] Mann, Vrij, and Bull, 2002.

[14] Oliver Lipps and Georg Lutz, "Gender of Interviewer Effects in a Multitopic Centralized CATI Panel Survey," *Methods, Data, Analyses*, Vol. 11, No. 1, 2017.

[15] Juan David Leongómez, Viktoria R. Mileva, Anthony C. Little, and S. Craig Roberts, "Perceived Differences in Social Status Between Speaker and Listener Affect the Speaker's Vocal Characteristics," *PLoS One*, Vol. 12, No. 6, June 14, 2017, e0179407.

[16] Hartwig and Bond, 2014; Hartwig and Bond, 2011; and Sporer and Schwandt, 2007.

[17] DePaulo et al., 2003.

[18] DePaulo et al., 2003.

[19] Sporer and Schwandt, 2007.

[20] Charles F. Bond Jr. and Bella M. DePaulo, "Accuracy of Deception Judgements," *Personality and Social Psychology Review*, Vol. 10, No. 3, 2006; Charles F. Bond Jr. and Bella M. DePaulo, "Individual Differences in Judging Deception: Accuracy and Bias," *Psychological Bulletin*, Vol. 134, No. 4, July 2008; and Hartwig and Bond, 2011.

In the following sections, we discuss three characteristics of deceptive speech (sentence complexity in responses, filler words, and deception in online communication) that some studies find are signals for detecting deception using a computational approach.

Sentence Complexity

Some studies find that deceptive statements tend to be less complex and include fewer details than truthful statements. One marker of sentence complexity is the use of *exclusive* or *exclusion words* (e.g., "but," "without," "exclude").[21] For instance, one study investigated lying behavior in five experiments that operationalized lying differently.[22] Using automated text analysis, the researchers found evidence across several studies that people who are telling the truth use exclusive words at higher rates than deceptive people do. Exclusive words might be less prevalent in deceptive speech because of "the cognitive load required to maintain a story that is contrary to experience, and the effort taken to try to convince someone else that something false is true."[23] Exclusive words are also often used to make category distinctions, another indication of complex speech.

Other markers of complex language shown to differentiate liars from truth-tellers include the use of the following language features:[24]

- conjunctions (e.g., "and," "also," "although")
- prepositions (e.g., "to," "with," "above")
- cognitive mechanisms (e.g., "cause," "know," "ought")
- words with greater than six letters.

For instance, the use of prepositions can signal that an interviewee is offering more-concrete information about a topic (and is therefore telling the truth).

Filler Words

Another linguistic feature that studies have shown to distinguish liars from truth-tellers is the use of so-called filler words (e.g., "um," "uh"). Increased usage of "um" is associated more often with truthfulness than with lies—both in laboratory-elicited,[25] low-stakes lies and in

---

[21] Yla R. Tausczik and James W. Pennebaker, "The Psychological Meaning of Words: LIWC and Computerized Text Analysis Methods," *Journal of Language and Social Psychology*, Vol. 29, No. 1, 2010.

[22] Matthew L. Newman, James W. Pennebaker, Diane S. Berry, and Jane M. Richards, "Lying Words: Predicting Deception from Linguistic Styles," *Personality and Social Psychology Bulletin*, Vol. 29, No. 5, 2003.

[23] Tausczik and Pennebaker, 2010, p. 34.

[24] Tausczik and Pennebaker, 2010; and Newman et al., 2003.

[25] Joanne Arciuli, David Mallard, and Gina Villar, "'Um, I Can Tell You're Lying': Linguistic Markers of Deception Versus Truth-Telling in Speech," *Applied Psycholinguistics*, Vol. 31, No. 3, July 2010; and Frank

real-life high-stakes lies.[26] For example, one study examined telephone and interview transcripts with Scott Lee Peterson (who was convicted of murdering his wife, Laci). By analyzing statements that could be verified as either truthful or deceptive, the researchers found that Peterson was more likely to use "um" in truthful rather than deceptive statements. This effect was observed in both informal telephone conversations and more-formal media interviews. Even among individuals who are skilled in impression management (i.e., TV personalities), "um" is used almost three times as often in a speaker's true statements than in their false ones.[27] Studies unrelated to deception have shown that speakers can reduce their usage of "um" via conscious control.[28] Therefore, it could be that liars are less likely to use filler words because they are strategically monitoring "their deceptive behaviors in [an] attempt to conceal 'leakage' of cues."[29]

Online Interactions

Because text messages and e-mail have become standard forms of communication, other linguistic cues for detecting deception have emerged from studying these interactions. For instance, deception involving the coordination and negotiation of social interactions, or *butler lies*,[30] are more common in text messages compared with face-to-face interactions because text messaging typically involves social coordination (e.g., arranging meeting times or planning other social interactions).[31] Examples of deception related to a person's "actions, whereabouts and plans" in texts include ambiguity about location (e.g., "I'm at the gym" when

---

Enos, Stefan Benus, Robin L. Cautin, Martin Graciarena, Julia Hirschberg, and Elizabeth Shriberg, "Human Detection of Deceptive Speech," 2006.

[26] Gina Villar, Joanne Arciuli, and David Mallard, "Use of 'Um' in the Deceptive Speech of a Convicted Murderer," *Applied Psycholinguistics*, Vol. 33 No. 1, January 2012.

[27] Gina Villar and Paola Castillo, "The Presence of 'Um' as a Marker of Truthfulness in the Speech of TV Personalities," *Psychiatry, Psychology, and Law*, Vol. 24, No. 4, 2017.

[28] Herbert H. Clark and Jean E. Fox Tree, "Using *Uh* and *Um* in Spontaneous Speaking," *Cognition*, Vol. 84, No. 1, May 2002; and Sabine Kowal, Daniel O'Connell, Katherine A. Forbush, Mark L. Higgins, Lindsay Clarke, and Karey D'Anna, "Interplay of Literacy and Orality in Inaugural Rhetoric." *Journal of Psycholinguistic Research*, Vol. 26, No. 1, 1997.

[29] Arciuli, Mallard, and Villar, 2010, p. 84

[30] Jeffrey T. Hancock, Jeremy Birnholtz, Natalya Bazarova, Jamie Guillory, Josh Perlin, and Barrett Amos, "Butler Lies: Awareness, Deception, and Design," *Proceedings of the SIGCHI Conference on Human Factors in Computing Systems*, April 2009.

[31] Madeline E. Smith, Jeffrey T. Hancock, Lindsay Reynolds, and Jeremy Birnholtz, "Everyday Deception or a Few Prolific Liars? The Prevalence of Lies in Text Messaging," *Computers in Human Behavior*, Vol. 41, No. C, 2014; and Agathe Battestini, Vidya Setlur, and Timothy Sohn, "A Large Scale Study of Text Messaging Use," *Proceedings of the 12th International Conference on Human Computer Interaction with Mobile Devices and Services*, September 2010.

actually at a bar) or social activities (e.g., "I'm hanging out with Person X" when actually one is with Person Y).[32]

In a study using a large dataset of text messages on Android smartphones, researchers found that untruthful text messages contained more words on average, particularly among students.[33] Students had more than a 24 percent increase in words per text when lying compared with students telling the truth. The study also found that women used about 13 percent more words per text when lying, compared with around 2 percent more for men. Liars engaging in interactive online conversations "may use more words to manage information flow, to enhance mutuality with their partner, and to decrease a conversational partner's suspicion."[34] They might also need more time to plan and edit their deceptive messages. Similarly, some users appeared to take longer and use more words to generate deceptive emails than they did truthful ones.[35]

Deceivers also have been shown to use more *noncommittal phrases* (e.g., *sure, possibly, probably*),[36] more *self-oriented words* (e.g., *I, I'm*), and fewer *other-oriented words* (e.g., *you, your*) when engaging in online text-based deception.[37] These patterns appear to vary by gender, with women using fewer other-oriented words while lying and men increasing their use of *my* and *me* (but using *I* significantly less). There also is evidence that these patterns vary with student status, with students using more noncommittal words, more self-oriented pronouns, and fewer other-oriented pronouns when lying in text messages than non-students do.

Machine Learning for Detecting Deceptive Speech

Recent developments in ML research have established that patterns of movement and language can reliably detect deception with better probability than chance. In this section, we summarize some of the recent findings from the literature and describe how our analysis fits into this existing landscape of results.

---

[32] Smith et al., 2014, p. 222.

[33] Jason Dou, Michelle Liu, Haaris Muneer, and Adam Schlussel "What Words Do We Use to Lie? Word Choice in Deceptive Messages," *arXiv*, 1710.00273, October 1, 2017; and Sarah Ita Levitan, Angel Maredia, and Julia Hirschberg, "Linguistic Cues to Deception and Perceived Deception in Interview Dialogues," *Proceedings of the 2018 Conference of the North America Chapter of the Association for Computational Linguistics: Human Language Technology*, Vol. 1, 2018.

[34] Hancock et al., 2008, p. 3.

[35] Lina Zhou, Judee K. Burgoon, Jay F. Nunamaker, Jr., and Doug Twitchell, "Automating Linguistics-Based Cues for Detecting Deception in Asynchronous Computer-Mediated Communications," *Group Decision and Negotiation*, Vol. 13, No. 1, 2004.

[36] Dou et al., 2017; and Levitan, Maredia, and Hirschberg, 2018.

[37] Dou et al., 2017.

Text-based ML methods to detect deception have some advantages over human perception methods or biological response methods (e.g., polygraphs). ML, by its nature, is truly data-driven: it does not have any preconceptions about which words or language features might be indicative of deception. Of course, this does not mean that bias cannot occur in ML analyses; indeed, we discuss some potential sources of bias in our own analysis in Chapter Five. Nevertheless, ML algorithms tend to work best with large amounts of data; this presents the researcher with some important tradeoffs. Large (i.e., $N > 1000$), high-quality datasets of interviews in which liars and truth-tellers are known are difficult to come by, so researchers tend to be forced to choose between small but high-quality datasets, and large but less-applicable datasets. Although we choose the first approach in this paper, the other approach shows some promise as well. Lai and Tan (2019), for example, observed that ML models could boost human accuracy at predicting deceptive hotel reviews, from about 50 percent to about 70 percent.[38] Deceptive text in other domains, including online dating and social networks,[39] also has been shown to be detectable via NLP methods. Pérez-Rosas and Mihalcea (2015) developed their own dataset of Internet user–contributed deceptive and truthful statements and found that lies could be detected in the best case about 69 percent of the time via an ML algorithm.[40]

Levitan et al. (2018) studied a text and voice recording dataset of university students in pairs asking each other personal questions about which they were instructed to lie half the time.[41] By training a Random Forest model on the words used in each response and language categories from the Linguistic Inquiry and Word Count (LIWC) database,[42] they found approximately a 70 percent accuracy rate (with comparable results for F1 score, precision, and recall). Yancheva and Rudzicz (2013) studied the transcripts of children called to testify in court cases and determined that various ML algorithms could achieve accuracies of around 65 percent at detecting deceptive speech and up to 90 percent for some small subgroups of children.[43] In short, a sizeable number of papers in the ML literature support the idea that a roughly 70 percent accuracy rate for discriminating between deceptive and truthful speech is possible via text and NLP analytic methods.

---

[38] Vivian Lai and Chenhao Tan, "On Human Predictions with Explanations and Predictions of Machine Learning Models: A Case Study on Deception Detection," *Proceedings of the Conference on Fairness, Accountability, and Transparency*, January 2019.

[39] Catalina L. Toma, Jeffrey T. Hancock, and Nicole B. Ellison. "Separating Fact from Fiction: An Examination of Deceptive Self-Presentation in Online Dating Profiles," *Personality and Social Psychology Bulletin*, Vol. 34, No. 8, August 2008; and Jalal S. Alowibdi, Ugo A. Buy, Philip S. Yu, Sohaib Ghani, and Mohamed Mokbel, "Deception Detection in Twitter," *Social Network Analysis and Mining*, Vol. 5, 2015.

[40] Verónica Pérez-Rosas and Rada Mihalcea, "Experiments in Open Domain Deception Detection," *Proceedings of the 2015 Conference on Empirical Methods in Natural Language Processing*, September 2015.

[41] Levitan, Maredia, and Hirschberg, 2018.

[42] For details about random forest, see IBM, "Random Forest," webpage, December 7, 2020.

[43] Maria Yancheva and Frank Rudzicz. "Automatic Detection of Deception in Child-Produced Speech Using Syntactic Complexity Features," *Proceedings of the 51st Annual Meeting of the Association for Computational Linguistics*, Vol. 1: *Long Papers*, August 2013.

Another line of research has been the study of nontext features of liars, particularly movement (e.g., hand gestures, facial micro-expressions), acoustical features, and meta-level language features. Khan et al. (2021),[44] for example, built a model with an accuracy rate approaching 80 percent by training it on facial and eye movements. Mendels et al. (2017)[45] examined the same dataset as in Levitan et al. (2018) that used deep neural networks to analyze the acoustic features of participants' voices, finding a maximum precision of about 76 percent.

Meta-level features, such as the total number of words used by an interviewee, have shown mixed results in the literature. DePaulo et al. (2003) found that liars speak more quickly than truth-tellers,[46] albeit using a similar total number of words. On the other hand, Levitan, Maredia, and Hirschberg found the opposite result: Liars spoke for a longer duration.[47]

Given that different techniques to detect deception each can be about 70 percent accurate, it is possible that a model that combines information from transcripts, movements, and audio cues could lead to higher performance than any individual technique. Such a model likely would require a large custom dataset that might be costly to produce but could be a promising avenue for future research. In this report, we combined several different data types (i.e., text transcripts, linguistic stances, meta-language features) to illustrate how such a system could be built. Our results illustrate the need for proper model regularization and sufficient data. Nevertheless, we show that combining multiple streams of text (from a chat interface and a video interview) is quite feasible and promising.

Summary

In summary, we have several key takeaways from this chapter. First, the evidence on deceptive cues is mixed. While some research indicates there are certain patterns in deceptive speech, other meta-analytic studies argue that cues to deception are weak at best. Second, research shows that individuals can display certain linguistic patterns when engaging in deceptive speech and these patterns have been detected in research using ML methods. Compared with truthful statements, deceptive statements usually are less complex, include fewer details, and are less likely to contain filler words. As online interactions have increased, researchers have also found that deceptive text-based speech is more likely to be related to a person's actions, whereabouts, and plans compared with lies told in face-to-face interactions. Untruthful text-based messages typically are also longer, take more time to write, and are more noncommittal

---

[44] Wasiq Khan, Keeley Crockett, James O'Shea, Abir Hussain, and Bilal M. Khan, "Deception in the Eyes of Deceiver: A Computer Vision and Machine Learning Based Automated Deception Detection," *Expert Systems with Applications*, Vol. 169, May 1, 2021.

[45] Gideon Mendels, Sarah Ita Levitan, Kai-Zhan Lee, and Julia Hirschberg. "Hybrid Acoustic-Lexical Deep Learning Approach for Deception Detection," *Proceedings of Interspeech 2017*, 2017.

[46] DePaulo et al., 2003.

[47] Levitan, Maredia, and Hirschberg, 2018.

and self-oriented than truthful text-based messages. Moreover, these patterns can vary based on the gender and status of an interviewee. In subsequent chapters, we describe ML models that have the capacity to detect some of these signals, specifically counts of commonly used words (word vectors), attitudinal dimensions of words (stance), and meta-level features of speech, such as speaking cadence (metadata). Results from these models show accuracy rates of between 62.2 and 75.6 percent in distinguishing truth tellers from people who were lying.

CHAPTER THREE

Description of Data

We recruited 103 participants associated with the University of New Haven in West Haven, Connecticut.[1] All participants were English speakers, U.S. citizens, and aged between 18 and 60. The study guaranteed all participants $50 for one and one-half hours of their time with the chance of earning up to $100 if the study staff decided they put forth sufficient effort during their interviews in this study. This was the same incentive for participants instructed to be truthful and for those who were instructed to be deceptive.[2] Unbeknown to all participants, the study paid everyone $100 for their participation regardless of effort or performance.

Sample and Procedures

Figure 3.1 displays the four-part order of operations for this study. First, participants arrived at our subcontractor's laboratory, signed an informed consent form, and completed a form self-reporting their demographic background information. Second, the subcontractor's study staff asked participants to read one of two accounts about Edward Snowden leaking classified information.[3]

For half of the participants, the story was presented as a news article with a fictitious newspaper logo prominently featured on the front page: "The Guardian: News You Need!"

---

[1] To confirm that this RAND report correctly described how the University of New Haven collected the experimental data, RAND requested that Dr. Morgan review and provide feedback in May 2022. Also, we note that the data we present and analyze in this RAND report are a subset of a broader set of experimental data collected by the University of New Haven as part of its subcontract with RAND. Although there was a single study protocol used in the collection of data, the experiments related to this overall effort were separate: One part of this design was led by RAND, and the other part of this design was led by the University of New Haven, each with distinct samples of participants who were randomly assigned to distinct experimental conditions. This report does not include results from the University of New Haven's part, which may be published in a non-RAND outlet in the future.

[2] Participants assigned to the truth-telling condition had to tell the truth only about what they had read. In comparison, participants in the lying condition had to omit key pieces of information about what they had read. Thus, participants in the lying condition likely had to put forth more effort during the study. This might have made it easier for our models to detect attempts at deception compared with those telling the truth.

[3] This fictitious article was based on a news report: Gregory Korte, "'I Have Done Nothing Wrong'," *USA Today*, June 10, 2013.

and a journalist listed as the author. For the second half of participants, the document was presented as a sensitive memo sent to a fictitious government official, "Ambassador Culvahouse." It was labeled "SENSITIVE DOCUMENT" at the top of the memo; the level of importance was listed as "HIGH" in red; the classification level was listed as "SENSITIVE" in red; and three paragraphs of text were in red with the fictitious classification marking "(S)" before each of these paragraphs (Appendix B displays these study materials).

FIGURE 3.1
Order of Operations for This Study

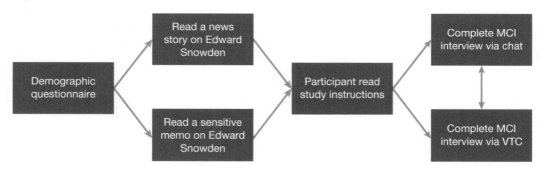

In the third part, the university's staff gave participants instructions on what they should say during a modified cognitive interviewing (MCI) session. For participants who read a news story about Edward Snowden, staff instructed participants to speak about the task that they had just completed when asked by interviewers. For participants who read the same document presented as a sensitive memo, the staff told participants not to reveal certain parts of what they had just read in this sensitive document and claim that they had read a news article.

In the fourth part, the study interviewed participants about what they had just read. All participants watched a short video by a former special agent from the Federal Bureau of Investigation (FBI) explaining the importance of being honest during this interview. Next, participants completed two versions of the same interview questions in random order: (1) a chat-based interview during which they answered questions about what they had just read via text-based messaging in real-time, and (2) a VTC interview with a trained interviewer in which they could see one another on a computer screen. After the interview, the study staff debriefed participants about the purpose of the study and paid them $100 for their time.

Table 3.1 displays the sample demographics for our sample. Our total sample size was 103 participants: 45 participants were assigned to the truth-telling condition (44 percent) and 58 participants were assigned to the lying condition (56 percent).[4]

---

[4] Table 3.1 shows that there was an additional $n = 13$ participants who were assigned to the lying condition. As a validation check, we re-ran our models after removing the last 13 people who were assigned to the lying

TABLE 3.1

Sample Demographics

Condition	Frequency	Percentage
Truth-Telling	45	43.7
Lying	58	56.3
Total	103	

Interview Technique

The interviewers used the same MCI method with all participants in the study (see Appendix A for more details about this interviewing technique), asking them a set of six detailed questions recounting what they did in the previous task.[5] What follows is a description of the script that interviewers trained in MCI used in this study.

At the beginning of the interview, participants were told by the interviewer:

> Well, as I said, my name is _____ and I am part of today's interview that is necessary before you can be awarded a security clearance. I am here to ask you questions about your life, your past, and about any activities that might be an issue for receiving a clearance. I've already gathered some information to support this investigation and now I am speaking with you so that I can move this process forward and clarify any issues or questions that I may have.

After completing the basic introductory phase of the interview, the investigator asked about specific activities that were linked to the task that the participant engaged in in the laboratory (i.e., reading of a news story or memo marked as sensitive).

> My investigation has shown that people, like you, often have contacts on the internet or have exposure to information from the internet that can be important to review. My investigation has also shown that you appear to have had some unusual activity, but I have not been able to understand what this is pertaining to. As you can see, this is a very serious matter, and so I must tell you it is important that everything you say during our interview is the absolute truth.

---

condition. The pattern of results did not substantially change from the results presented in Chapters Four and Five.

[5] Charles A. Morgan, Yaron G. Rabinowitz, Deborah Hilts, Craig E. Weller, and Cladimir Coric, "Efficacy of Modified Cognitive Interviewing, Compared to Human Judgments in Detecting Deception Related to Bio-Threat Activities," *Journal of Strategic Security*, Vol. 6, No. 3, Fall 2013; and Charles A. Morgan, III, Yaron Rabinowitz, Robert Leidy, and Vladimir Coric, "Efficacy of Combining Interview Techniques in Detecting Deception Related to Bio-Threat Issues," *Behavioral Sciences and the Law*, Vol. 32, No. 3, May–June 2014.

I'd now like to ask you some questions about your activities on the computer; specifically, I also have to ask you some questions about whether you have ever been exposed to or mishandled sensitive information.

These trained interviewers then asked participants: (1) to fully recall what happened when they read about Edward Snowden's leaking of classified information, (2) to recall the visual details during this activity, (3) to recall the auditory details during this activity, (4) to recall how they felt during this activity, (5) to respond to a temporal recall question about this activity, and (6) to disclose information they might have left out about this activity. The following is the script that these interviewers followed:

1. First ask a "full recall" prompt, designed to have the participant recount in detail their memory of their time when on the computer prior to the interview:
   – Please tell me everything you remember about the materials that you viewed before you came to meet with me.
2. A sensory prompt designed to elicit visual details:
   – Now, I'd like you to imagine that I had been with you while you were reading the paper. But imagine that I was deaf and couldn't hear anything. I can only watch it like it is a silent movie. What would I have seen during that time?
3. A sensory prompt designed to elicit auditory details:
   – Now imagine I am there during this time, and I can hear, but I can't see, as if I was a blind person. What would I have heard during this time while you were looking at the papers?"
4. An affective prompt designed to elicit details regarding the participant's emotional state of mind during their time reading about Edward Snowden leaking classified information:
   – How was that experience? What were you feeling during that time?
5. A temporal prompt in which the interviewee was asked to recount what happened during their time reading about Edward Snowden leaking classified information by starting with the last thing that they recalled happening and ending with the first thing they remember happening:
   – Now I'd like you to start with the very last thing you "remember" (for example, closing the computer). I'd like you to start with the last thing that happened and walk me backwards through your memory of what you remember. So if the last thing that you remember was you walking out the door to come and see me, what do you remember right before that?"
6. A final prompt in which the participant was afforded the opportunity to consider whether they had left anything out of their recounting of their time reading about Edward Snowden leaking classified information or whether they wanted to change or correct anything about their recollection:
   – In telling me what you remember, do you think you left anything out? Do you think you made any mistakes?

Analysis of Interview Data

We used two sources of data. First, our subcontractor audio recorded all interviews with the consent of participants. To transcribe the files, researchers from RAND uploaded these recordings to Amazon Web Services (AWS) Transcribe. AWS Transcribe uses a deep learning process—automatic speech recognition—to convert the recordings from speech to text.[6] AWS Transcribe's automatic speech recognition technology also identifies the responses by interviewers and interviewees. (An example of the raw output from AWS Transcribe is displayed in Appendix C.) Next, a team of research staff listened to each audio recording and manually checked the AWS Transcribe outputs for each interviewee and revised mistakes.[7] After completing these corrections, the team combined the files for analysis.

The second source of data was text logs of the virtual chats that all interviewers completed either before or after their VTC interviews. The questions asked during the chats and VTC interviews were identical. Members of the subcontractor who collected these data compiled these chat exchanges in a Microsoft Word document, and then RAND research staff used these files to conduct a separate analysis of the text files.

Analysis Focused on Detectable Differences During VTC and Chat Interviews

The outcome measures of this study are what participants said during their VTC interviews and what they typed when asked the same questions during the chat interview. As an exploratory study, our focus is on whether there are detectable differences in how people respond to the same questions—using the same interview technique—when they are trying to deceive a trained interviewer versus when they are being honest. Chapter Four reviews the results from this exploratory analysis using several NLP models.

---

[6] Amazon Web Services, "Amazon Transcribe," webpage, undated.

[7] AWS Transcribe frequently missed filler words that participants expressed (e.g., *um* or *uh*), which our research team had to manually correct.

Results from Analysis of Interview Data

In this chapter, we present results from the exploratory models that we developed to detect attempts at deception versus those who were being truthful during interviews conducted by VTC and virtual chat. We developed several models that detect differences in what was said—and how it was said—among participants trying to be truthful or deceitful. This chapter has four parts. First, we present the background for our linguistic models that used word counts, stance, patterns throughout interviews (which we call metadata), and a combination of these models. Second, we present results from these models for transcripts from interviews conducted via VTC. Third, we present results from these models for chat-based interviews. Finally, we summarize the key findings from this exploratory research.

Background of the Linguistic Models

Because there is no one-size-fits-all approach to analyzing unstructured text, such as the transcript of an interview, we built several classes of models to identify deceptive speech, focusing on different characteristics of the interview data. The first model analyzed differences in the types and frequencies of words that participants used during their interviews (for example, the frequency of the words *truth* or *um*). The second model looked at the linguistic stance of the text: the presence of characteristics such as emotion or doubt. The third model used metadata from the interviewer-interviewee interactions, such as average word length and the fraction of words used with more than six letters. We also explored models that analyzed various combinations of the VTC and chat datasets, which (in some cases) outperformed the individual models. Appendix D describes an attempt to build a proof of concept for a separate model, a deep neural network model that is trained to identify contradictions in pairs of statements.[1]

---

[1] This contradiction model is intended to catch interviewees in a lie when follow-up questions make clear that a particular response is inaccurate.

Technical Overview of Models

Our first three models were built around a relatively straightforward ML exercise: When given the transcripts of interviews with truth-tellers and liars, can an algorithm perform binary classification more accurately than chance? And if so, could we interrogate that model to understand how it made its decisions?

Binary classification is a well-worn problem in ML, and mature literature exists on how to train classifier algorithms and validate their results. Implementations in Python are readily available through several ML packages. We used Scikit-learn, version 0.23.2, in this work.

Classification algorithms typically require clean, numeric data to perform well. The key step for building our binary classifier, then, was feature-building: translating the unstructured text of interview responses into numerical values. We used three different methods to build our numerical features, which are described in the following paragraphs. Each model is identical except for the feature vectors x. The feature vectors were fed directly into a logistic regression classifier, which models an outcome y_i as a Bernoulli (binary) variable given by the probability:

$$\theta_i = \frac{e^{\beta \cdot x}}{1 + e^{\beta \cdot x}}.$$

Here, x is the feature vector and β is the vector of coefficients that are optimized by maximizing the likelihood:

$$L(\beta) = \prod_i \theta_i^{y_i} (1 - \theta_i)^{1 - y_i}.$$

As is standard practice in ML applications, we impose an L2 regularization scheme on the vector β, and for numerical simplicity we use the log-likelihood. Therefore, the model actually is trained by maximizing the modified log-likelihood:

$$\log L(\beta) = \sum_i \left\{ (\beta \cdot x) y_i - \log(1 + e^{\beta \cdot x}) \right\} + \frac{C}{2} \sum_i \beta_i^2.$$

Our cross-validation scheme uses three different regularization constants for C, with values of 5, 10, and 20. Still, with a relatively small ($n = 103$) number of labeled samples to work with and a comparable or larger number of features (500, in this case) to fit, we sought to avoid overfitting the data. Overfitting occurs when a model performs well on the data it is trained on but generalizes poorly to unseen data. In our case, our word vectorization model has more parameters than the number of samples in the training dataset, meaning that it is easy for the model to overfit.

We attempted to mitigate other sources of overfitting by using a nested three-fold cross validation when evaluating our models. Cross-validation is a common generalization of the standard test/train splitting used in ML approaches, in which a subset of data is withheld from training and used to simulate unseen data when evaluating the model performance. In the three-fold cross-validation strategy, we split the data into training and testing sets three times, forming nonoverlapping test sets each time, and we train three independent models on the associated training sets. The mean performance across the test sets is then a better approximation for out-of-sample performance than a single test/train evaluation (insofar as the dataset is representative of real-world data, at least) because any statistical fluctuations from generating a small test set are smoothed out.

However, cross-validation alone might create another potential reason for overfitting: The hyperparameters of the models could be uniquely suited to the training data used when selecting them.[2] The *nested* cross-validation that we used tackles this by combining a grid search for hyperparameters with cross-validation, yielding the best model and a relatively unbiased estimate for its accuracy. The available literature suggests that applying nested cross-validation in this way can yield robust models even when the sample size of a training set is small compared with the model complexity.

No amount of careful training protocol can overcome potential sources of systematic biases of the training data that we used: Our transcripts were collected in a controlled environment, with low stakes for people caught lying, and real-world conditions might differ in some way that would make our performance metrics overly optimistic or pessimistic. In the absence of more-comprehensive data, however, we believe that our results are as correct as possible. We describe these and other limitations later in the report.

Analysis of Virtual Teleconference Interviews

This section describes the models used to analyze participants' responses to the trained interviewers conducted via VTC.

Model 1: Word Vectors

The first model that we built is simple: We simply counted the number of times that our interviewees used common words, a method called *word count vectorization*. Specifically, our vectorizer counted the number of occurrences of each of the 500 words used most across all interviews. For example, the sentence "The researcher wrote the report" would have a corresponding vector of $x = [2, 1, 1, 1]$ with the associated dictionary {"the", "researcher", "wrote", "report"}. In this case, the word *dictionary* simply refers to an ordered list of unique words that might appear in a sample of text.

Note that this implementation is not normalized; an interviewee who speaks for longer will tend to end up with higher counts than an interviewee who is more taciturn. Still, many words are not used by any individual interviewee, so the word count vector for each entry is somewhat sparse. Also note that we compute the word vectors for the combined responses of each interviewee, not for each response separately.

After training each of our models, we studied them to analyze the key features that allowed them to distinguish between liars and truth-tellers. We did this by examining the relative

---

[2] *Hyperparameters* are model parameters that are not automatically learned during training and must be specified by the developer.

importance of different words and bigrams to the model, both by performing a variable-importance study and by computing which words were used most differently by liars and truth-tellers.

Specifically, we shuffled the number of usages for each word, retrained the model on the permuted dataset, and measured the decrease in performance compared with the baseline. The idea behind this method is that unimportant words, when shuffled, will have little effect on the model, but scrambling the highly important words will make the model perform significantly worse. Importance is computed via a permutation algorithm in which the values for each word count are scrambled among the interviewees, and the accuracy of the classifier on the resulting dataset is compared with the full dataset.

Figure 4.1 displays the highest word importance scores computed by this permutation method for Model 1 in predicting the accuracy of which participants were truthful or lying in our study. Words more commonly used by liars are in red, while words more commonly used by truth-tellers are in blue. Figure 4.1 shows that the words *I* and *you* were features that significantly improved our models' accuracy at predicting lying or truth-telling. Across all our models, autobiographical statements (use of *I*) are much more common for truth-tellers, while liars appear to be far more likely to deflect attention away from themselves with the

FIGURE 4.1
Feature Importance, Model 1

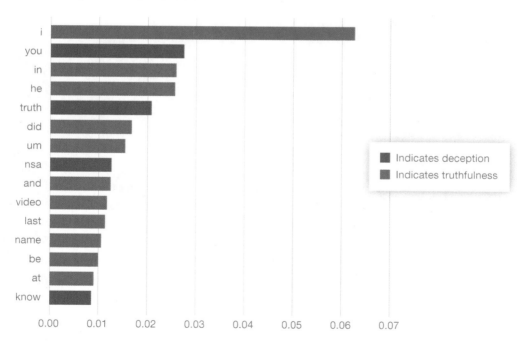

Feature importance (accuracy change)

word *you*. We note that other studies have found the opposite result, with second-person pronouns used less by deceivers than truth-tellers.[3]

The figure shows that using the word *I* increased the accuracy of predicting lying from
truthfulness by about five percentage points. Furthermore, the words *you* and *truth* increased
the accuracy in our models by about 2 and 1.5 percentage points, respectively. While these
changes are small on an absolute basis, Model 1 considers 500 different words in its feature
vector, so even a 1 percentage point change for a single word is substantial.

With these results to guide us, we looked at the frequency of the words *I*, *you*, and *truth*
used by the participants in our study. Figure 4.2 shows that people trying to lie used the word
I 30.8 times, on average, during their interviews, versus 41.2 times for truth-tellers. This difference was statistically significant. Furthermore, liars used the word *you* 12.7 times, on average, versus 5.0 times for truth-tellers. That difference was also statistically significant.

FIGURE 4.2
Average Frequency of Select Word Uses by Study Condition

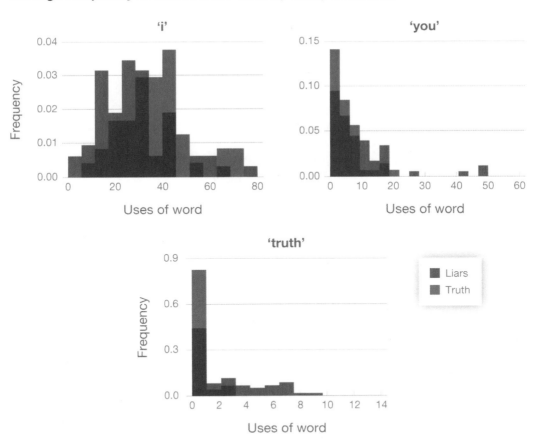

3 Bob de Ruiter and George Kachergis, "The Mafiascum Dataset: A Large Text Corpus for Deception
Detection," *arXiv*, 1811.07851, last revised August 14, 2019.

Finally, the word *truth* was used, on average, 0.33 times by truth-tellers versus 2.8 times by those trying to lie. That difference was statistically significant as well. Note that *truth* generally was used in the context of the interview (participants were asked to watch a video on the importance of telling the truth), as opposed to using a phrase like *to tell the truth*, but we note that liars still used the word significantly more frequently than truth-tellers. For the words *you* and *truth*, there was significant overlap at low usage rates among liars and truth-tellers; a lie-detector scheme based around just those words therefore is likely to have a low false-positive rate, albeit at the cost of a high false-negative rate.

Model 2: Stance Vectors

Our second model tested the use of *stance* vectors computed by the RAND-Lex suite of language processing tools.[4] Here, stance is an extension of sentiment to include other attitudinal dimensions of the language such as certainty or social relation. RAND-Lex has access to a proprietary dictionary of words and phrases that are binned into 119 stance categories. For example, the word *later* might be categorized into the "Time Shift" stance, while the word *said* might be categorized under the "Communicator" stance. The result of this processing for a piece of text is a vector length of 119 whose entries are a real number between zero and one, corresponding to the fraction of the text that falls into each stance category. The stance vectors are relatively sparse (for a short piece of text, typically between 70 and 90 percent of the stances have a value of zero), although not as sparse as the word count vectors from Model 1.

Because the RAND-Lex stance dictionaries are relatively content-agnostic, we expected that they would provide a more durable set of features than simple word counts, which are liable to change if a different topic is being discussed. The RAND-Lex stances also are readily interpretable. On the other hand, the RAND-Lex dictionaries are not exhaustive of all words in English, and they bin many different words into the same category. The resulting vectors therefore contain somewhat less information compared with a word-counting approach. It was unclear at the study's outset whether the advantages of a RAND-Lex approach would

---

[4] Our stance model uses a taxonomy of variables originally developed at Carnegie Mellon University to capture rhetorical and pragmatic effects in text. For references, see Hannah Ringler, Beata Beigman Klebanov, and David Kaufer, "Placing Writing Tasks in Local and Global Contexts: The Case of Argumentative Writing," *Journal of Writing Analytics*, Vol. 2, 2018; and Danielle Wetzel, David Brown, Necia Werner, Suguru Ishizaki, and David Kaufer, "Computer-Assisted Rhetorical Analysis: Instructional Design and Formative Assessment Using DocuScope," *Journal of Writing Analytics*, Vol. 5, 2021. RAND-Lex is RAND's proprietary text analysis software platform. For examples of research that use RAND-Lex, see: William Marcellino, Christian Johnson, Marek N. Posard, and Todd C. Helmus, *Foreign Interference in the 2020 Election: Tools for Detecting Online Election Interference*, Santa Monica, Calif.: RAND Corporation, RR-A704-2, 2020; and William Marcellino, Todd C. Helmus, Joshua Kerrigan, Hilary Reininger, Rouslan I. Karimov, and Rebecca Ann Lawrence, *Detecting Conspiracy Theories on Social Media: Improving Machine Learning to Detect and Understand Online Conspiracy Theories*, Santa Monica, Calif.: RAND Corporation, RR-A676-1, 2021.

outweigh the drawbacks for this application. Recent RAND work has shown that the inclusion of stance features often can aid in performance and interpretability of linguistic models.[5]

We computed the biggest differences in stance usage between liars and truth-tellers. This was done by taking the mean value of each stance across the liars and subtracting the mean value of each stance across the truth-tellers, to arrive at what we call "over-present" stance vectors. Table 4.1 displays the top over-present stance vectors by ranking the stance categories by differences in stance value. For truth-tellers, the top stances identified using RAND-Lex were autobiographical statements (e.g., *I used to*), linguistic references (e.g., *noun, verb*), time and date (e.g., *July 4, 1776*), common authorities (e.g., *the courts*), and anger (e.g., *upset*).

In comparison, the top stance vectors identified for those in the deceptive condition were references to you/attention (e.g., *look*), uncertainty (e.g., *maybe*), projecting back (e.g., *used to*), subjective perception (e.g., *it seems*), and oral cues (e.g., *you guys*). The uncertainty stance is a particularly good test of the hypothesis that lying places a mental strain on interviewees. Figure 4.3 shows that the uncertainty stance was used more by liars than by truth-tellers.

Figure 4.3 displays comparisons for three of the stance vectors by the liar or truth-teller conditions assigned to participants. The distributions of all three stances differ in a statistically significant way between liars and truth-tellers. These results align with our findings, shown in Figure 4.1, that the words *I* and *you* were features that significantly improved our models' accuracy at predicting lying from truth-telling. Across our models for interviews done via VTC, autobiographical statements (e.g., use of the word *I*) are much more common for truth-tellers, while those trying to lie appear to be more likely to deflect attention away from themselves with the word *you*.

TABLE 4.1

Top Over-Present Stance Vectors by Study Condition

Rank	Truthful	Deceptive
1	Autobiography	You/attention
2	Linguistic reference	Uncertainty
3	Time and date	Projecting back
4	Common authorities	Subjective perception
5	Anger	Oral cues

---

[5] Christian Johnson and William Marcellino, *Bag-of-Words Algorithms Can Supplement Transformer Sequence Classification and Improve Model Interpretability*, Santa Monica, Calif.: RAND Corporation, WR-A1719-1, 2022.

FIGURE 4.3

Comparison of Select Stance Vectors by Study Condition

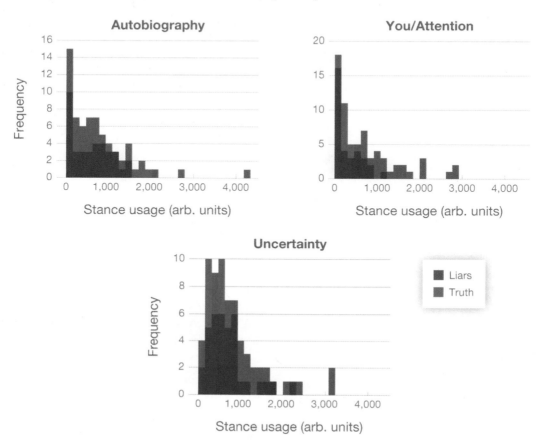

NOTE: This figure displays stance usage, in arbitrary units, for liars and truth-tellers for three different stance categories. Stance values in each category (a real number between 0 and 1) are taken from RAND-Lex and accumulated across each statement by interviewee and then weighted by the number of words that each interviewee uses. Therefore, the exact values are not straightforward to interpret, but the trend across our two categories of truth-tellers and liars is distinct.

Model 3: Metadata

Our third model examined other, meta-level features of speech, such as speaking cadence. We also considered five other types of features: total number of words used, number of unique words used, average word length, average unique word length, and fraction of words with more than six letters. (Our feature vector was therefore given by x = [cadence, total words, number of unique words, mean word length, mean unique word length, and fraction of words with more than six letters]). We compared the distribution of these values among liars and truth-tellers by plotting histograms (see Figures 4.4 through 4.6).

We compare each variable's distribution between the lying and truth-telling populations by performing a Kolmogorov-Smirnov test to compute a two-tailed p-value (which is interpreted as the likelihood that the liar and truth-teller values were drawn from the same distribution).[6] We considered a p-value at or less than 0.05 to be statistically significant.

None of the "meta" features were, by themselves, found to be statistically significant ($p > 0.05$ in all cases). Nevertheless, several of the p-values derived from our Kolmogorov-Smirnov tests were only slightly higher than 0.05, in agreement with what can be seen in the distributions shown in Figures 4.4 through 4.6: Liars and truth-tellers appear to differ very slightly in these meta speech features, even if the distinctions are not individually meaningful. We hypothesized that these individually non-significant features, when combined, would lead to an overall classifier with good performance. Much like with the stance data, we hypothesized that these features would be more durable even if the topics of conversation changed significantly.

Figure 4.4 displays the average total word length for participants in the truth-teller and liar conditions (on the left side) and the average number of unique words identified during these interviews (right side). The average word length among truth-tellers was 3.95 letters versus 3.96 letters among the liars. These differences were not statistically significant. On the right side of Figure 4.4, we see the average unique word length for the truth-tellers was 4.73 letters versus 4.63 letters for the liars. These differences were also not statistically significant ($p = 0.053$).

Figure 4.5 looks at the average words per second used by participants (on the left side) and the fraction of words that were under six letters (on the right). The left side shows that the

FIGURE 4.4

Word Lengths Used by People in Truthful and Deceitful Conditions

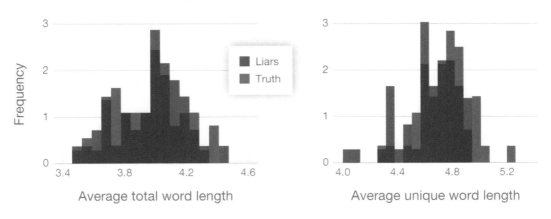

NOTE: The left panel shows the average length of the words used in number of letters.

---

[6] For more details, see National Institute of Standards and Technology, *NIST/SEMATECH e-Handbook of Statistical Methods*, Gaithersburg, Md.: U.S. Department of Commerce, last updated April 2012; and National Institute of Standards and Technology, "1.3.5.16, Kolmogorov-Smirnov Goodness-of-Fit Test," *NIST/SEMATECH e-Handbook of Statistical Methods*, Gaithersburg, Md.: U.S. Department of Commerce, last updated April 2012.

FIGURE 4.5

Words per Second and Fraction of Smaller Words for Truthful and Deceitful Speech

liars expressed, on average, 2.26 words per second versus 2.11 for the truth-tellers. This difference is not statistically significant ($p = 0.067$). The right side shows that 15.5 percent of the words expressed by truth-tellers were six letters or more versus 14.8 percent for liars. These differences are also not statistically significant ($p = 0.091$).

The result that liars tend to speak slightly more quickly than truth-tellers is at odds with our initial hypothesis that liars would speak more slowly to choose their words more carefully. This could be explained by the result in the right panel (the fraction of words used that are greater than six letters). The literature suggests that this metric differs significantly between deceptive and truthful speech, and we do see a difference in our sample. Liars tend to use fewer complex words, which aligns with our initial hypothesis that deception requires extra mental effort that detracts from the use of complex speech. With fewer complex words, liars might have had an easier time achieving a faster speaking cadence. Note that, in both cases, the distributions, while distinct, are not different enough to singularly identify any individual as a liar.

Figure 4.6 shows the total number of words and number of unique words that participants expressed. The left side of the figure shows that, on average, liars used 705 words during their interviews versus 725 words for truth-tellers. These differences are not statistically significant. The right side shows that liars on average expressed 164 unique words versus 179 words for truth-tellers. These differences are also not statistically significant.

FIGURE 4.6

Number of Total Words and Unique Words Used by People in Liar and Truth-Telling Conditions

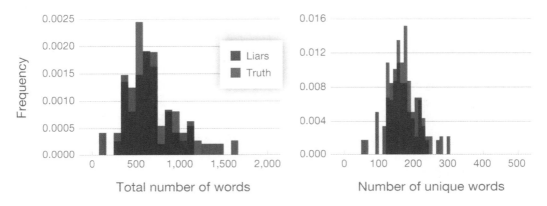

Comparison of Model Accuracy Rates

All of our binary classification models consistently performed better than chance at distinguishing between deceptive and truthful speech, although performance varied from model to model. The accuracy rates and performance metrics for the best-performing models are shown in Table 4.2, along with the results for the metadata, stance, and combined models.

We found that Model 1, which looked at the word vectors derived from the interview transcripts, was accurate 75.6 percent of the time—the best-performing model that we developed. Model 2, which classified the transcripts using the stance features of their language, correctly predicted which participants were being truthful or deceptive 64.1 percent of the time. Finally, our metadata model (Model 3), which was trained on derived patterns in what participants said throughout the entire interview, performed similarly to Model 2, with 62.7 percent accuracy.

Table 4.2 also shows the results of four combined models that used multiple factors together to predict liars from truth-tellers. By a *combined model*, we mean that the model was trained on the feature vectors created via direct concatenation of the feature vectors associated with Models 1, 2, and 3. Using the aforementioned examples, the combined Models 1 and 3 would produce the following feature vector for the sentence "the researcher wrote the report":

$$x = [2, 1, 1, 1, 3.5, 5, 4, 5.4, 6, 0. 2].$$

The first four entries correspond to the same x as before for Model 1, and the remaining six entries correspond to cadence, total number of words, number of unique words, mean word length, mean unique word length, and fraction of words with more than six letters for the sentence.

Combined Model 1, using word vectors and metadata, was accurate 74.5 percent of the time—our best-performing combined model. Combined Model 2, using word counts and stance, was 63.2 percent accurate. Similarly, Combined Model 3, using stance and metadata,

TABLE 4.2

Classification Performance of Exploratory Linguistic Models

Model Classification	Accuracy Rate (percentage)
Modeling by Factors	
Model 1: Word counts	75.6
Model 2: Stance	64.1
Model 3: Metadata	62.7
Modeling That Combines Factors	
Combined model 1: Word counts and metadata	74.5
Combined model 2: Word counts and stance	63.2
Combined model 3: Stance and metadata	62.2
Combined model 4: Word count, stance, and metadata	62.2

was 62.2 percent accurate. Combined Model 4 combined word counts, stance, and metadata for a 62.2 percent accuracy rate. We hypothesize the reduction in accuracy is because of our model regularization scheme (based on an L2 norm), which disproportionately penalizes models that have high numbers of parameters to prevent overfitting. Different regularization methods might produce different results, although we leave this question to future work. Because the algorithms theoretically should overfit more easily with more features, we consider this an indication that the regularization and nested cross-fold validation successfully penalized these models. Under this hypothesis, we would expect these combined models to outperform the others with a sufficiently large sample size.

We present our main results simply as accuracy rates; given the relatively balanced sample that we have and the cross-validation scheme that we used, this is appropriate.[7] Our accuracy rates are taken as an average across several models, meaning there is no single confusion matrix to report. However, for completeness, we computed the confusion matrix in Table 4.3 for a representative iteration of our best-performing model (Model 1). Given a random 80/20 train/test split, we found evidence that the model was overfit (which is why we performed nested cross-validation). However, accuracy on the test set is still reasonably well at 79 percent, which compares favorably with the overall accuracy we report in Table 4.3. We also found nothing significant about the ratio of false positives and false negatives.

TABLE 4.3
Representative Confusion Matrix for Model 1

	Train			Test	
	Actually True	**Actually False**		**Actually True**	**Actually False**
Predicted True	42	0	Predicted True	7	3
Predicted False	0	34	Predicted False	1	8

---

[7] Removing a handful of samples from the data to make them completely balanced had little to no effect on any of the major results.

Analysis of Virtual Chat Interviews

This section applies the same models that we discussed previously to transcripts of text-based chats that participants completed. All participants in this study completed the same interview via chat or VTC in a random order. We present the results of our models for detecting deception in these chat logs, and then summarize our key results from these models.

Methods

Models 1 (word vectors), 2 (stance), and 3 (word counts and stance) were trained in the same manner as described previously for the VTC interviews. The only new model in this chapter is Model 4, which combines (concatenates) the text transcripts of the VTC interviews and chat logs for each interviewee. As with the VTC interviews, we computed the word count vectors for the combined text transcripts and trained a logistic regression classifier. We did not use a metadata model because some previously used variables (e.g., cadence) do not apply to the chat logs, so an apples-to-apples comparison with the VTC interviews was not possible.

Results

We found that the chat logs generally provided a similar level of discriminating power as the VTC transcripts for each of our models. As stated earlier, participants completed the same interview, using MCI, via VTC with a live interviewer or via chat. Our subcontractor randomized the order in which participants completed both interviews. Some interviewees completed the VTC interview first, followed by the chat interview; others completed these interviews in reverse.

Table 4.4 shows the performance of our models using the chat logs. Model 1 used analysis of word vectors, which produced an accuracy rate of 65.6 percent. Model 2 used stance, while Model 3 used word counts with stance, both of which produced accuracy rates of 59.7 percent. Finally, for each interviewee, we used our word count model on their chat and VTC interview data, which produced the highest accuracy rate of 71.8 percent.

TABLE 4.4

Classification Performance of Models on Chat Logs

Model	Accuracy Rate (percentage)
Model 1: Word counts	65.6
Model 2: Stance	59.7
Model 3: Word counts and stance	59.7
Model 4: Chat and VTC word counts	71.8

NOTE: This table presents the accuracy scores for the different models that we developed, broken down by the two types of interviews conducted (informational and financial). The highest accuracy is given by the simple word-counting model, and we see little improvement when combined with metadata. Stance appears to be a relatively poor predictor of deceptive speech in general.

Figure 4.7 is analogous to Figure 4.6 (displayed earlier in this chapter), except that it shows the data for the combined transcripts of chat and VTC interviews. Similar to the VTC data alone, we see little difference in the distributions between liars and truth-tellers, which suggests that overall loquaciousness is not nearly as indicative of deception as the particular choice of words used.

FIGURE 4.7

Number of Total Words and Unique Words Used by People in Liar and Truth-Telling Conditions, Chat Logs

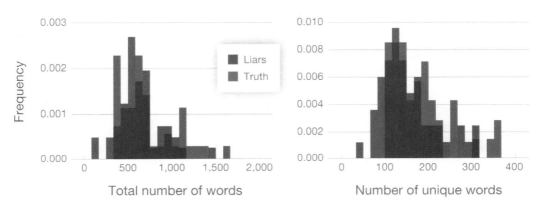

Next, we studied feature importance (using the same permutation method discussed earlier in this chapter) for the chat logs. Figure 4.8 shows the top over-present stance features in the two data sets, calculated the same way as the earlier VTC interviews. Similar to our earlier analysis, the word *you* was a strong predictor of deception in our analysis of the chat logs.

FIGURE 4.8

Word Importance for Combined Chat and VTC Model

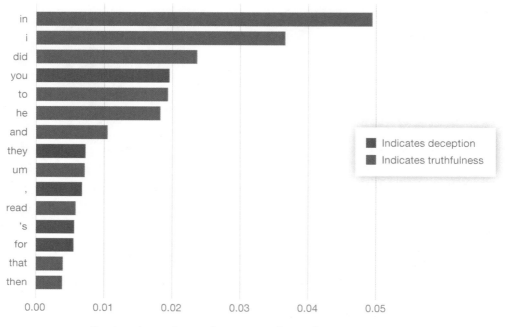

Feature importance (accuracy change)

NOTE: This figure shows a cascading plot of word importance to our best-performing chat and VTC model. Compared with Figure 4.1, we see overlaps (such as the words *I* and *you*), which makes sense because the model is looking at much of the same data.

Figure 4.9 shows the frequencies of the words "you," "I", and "in." This figure shows that the liars used the word "you" an average of 16.8 times during the interviews versus 7.8 times for truth-tellers. Furthermore, this figure shows that liars used "I" 44.4 times versus 57.2 times for truth-tellers. Finally, liars used "in" 13.2 times versus 18.3 times for truth-tellers.

FIGURE 4.9

Average Frequency of Select Words Used by Study Condition, Chat Logs

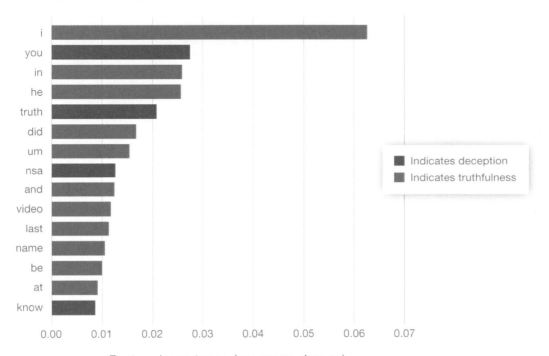

Feature importance (accuracy change)

Table 4.5 shows the biggest differences in stance usage between liars and truth-tellers. As described earlier, we calculated this by taking the mean values of each stance across the liars and subtracting the mean value of stance across the truth-tellers.

Table 4.5 also displays the top over-present stance vectors. In this case, self-disclosure (e.g., *I think*), citations (e.g., *according to*), projecting ahead (e.g., *in order to*), and person/ pronoun (e.g., *he, she*) are newly over-present among truth-tellers, while generic events (e.g., *working*), curiosity (e.g., *what should*), and motions (e.g., *run*) are new entries among the liars.

TABLE 4.5
Top Over-Present Stance Vectors by Study Condition, Chat Logs

Rank	Truthful	Deceitful
1	Self-disclosure	You/attention
2	Language reference	Project back
3	Citations	Curiosity
4	Project ahead	Generic events
5	Person/pronoun	Motions

As shown in Figure 4.10, we again see the same broad stance patterns emerge from the chat logs as for the VTC interviews: for liars, *you*-based language is more prevalent, while for truth-tellers, *I*-based language is more common. This might explain why the combined model (Model 4) performs well with 71.8 percent accuracy: It is combining these signals, which improves any signal-to-noise ratio. This suggests that the more text that can be gathered in an interview setting, the better this class of model will perform. Engaging the interviewees in a longer, back-and-forth chat will be more productive than asking just a few questions, and better still is to combine the data across different interview methods, which will help alleviate the potential effects of bias that may be present in one medium. We discuss one such possible source of bias in Chapter Five.

FIGURE 4.10

Comparison of Select Stance Vectors by Study Condition, Chat Logs

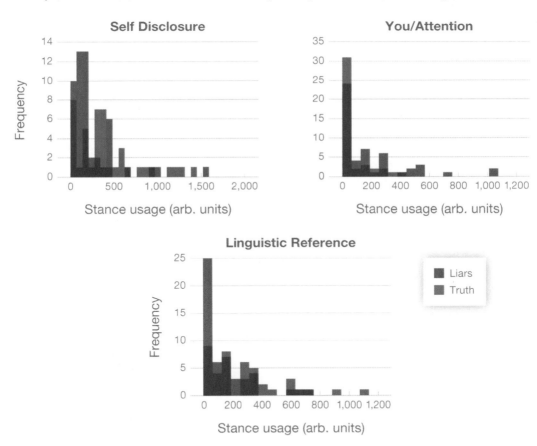

NOTE: Similar to Figure 4.3, we show the usage of different stance features for the chat log data. We find a similar pattern as before: While there is considerable overlap of each distribution, the tails of each distribution are relatively distinct and can be helpful for identifying deceptive or truthful speech. We see a similar pattern in the use of first-person and second-person language, with the former more associated with truth-telling and the latter with deception.

Summary of Key Findings

This chapter described exploratory models that used different factors to predict who was trying to be truthful or deceptive during our study. We developed three models that used different factors to make this prediction. Model 1 used word vectors, Model 2 used stance vectors, and Model 3 used *metadata*, defined as patterns of responses that individuals gave during their interviews. All of these models, for both VTC and chat logs, produced results that were better than random chance, with accuracy rates ranging from 62.7 to 75.6 percent. Model 1, which used total word counts and the number of unique words to predict who was being truthful or lying, was the most accurate.

We then combined these factors into new models. The accuracy rates in predicting liars and truth-tellers from these models ranged from 62.2 to 74.5 percent. The best-performing of these models (Combined Model 1) combined word counts and metadata. We also found that combining Model 1 for chat and VTC interviews produced an accuracy rate of 71.8 percent.

CHAPTER FIVE

Potential Sources of Bias

This chapter provides an overview of potential sources of bias based on the gender of interviewees in the results from our models. There is some evidence that men and women tend to deceive in different ways.[1] Accordingly, our team explored whether deceptive speech among the interviewees differed by gender, and if so, whether the efficacy of our models might also vary by gender as a result.

We draw from the literature on gender and deception to guide our analyses in what was an iterative process. We used the same methods for the VTC chats that we described in the previous chapter. However, for these data, we analyzed results for men and women separately to determine whether model accuracy rates significantly differed by gender. In the following section, we provide a brief overview of the literature and describe our findings. Next, we offer possible explanations for our results. We conclude with implications and other considerations regarding the potential for bias when using algorithmic tools for detecting deception.

Research Finds Gender Might Affect How People Are Deceptive

The relationship between gender and deceptive behavior is well documented. Although results are somewhat mixed regarding the role that gender plays in the likelihood of telling a lie, there is some evidence that gender may affect *how* people tell lies. Men and women may use different words when they lie,[2] and they might have different reasons for lying as well.[3] Although the evidence is mixed, both longitudinal and experimental studies have found that men and women could be motivated to lie for different reasons. Some research indicates that

---

[1] Dou et al., 2017; Anna Dreber and Magnus Johannesson, "Gender Differences in Deception," *Economics Letters*, Vol. 99, No. 1, April 2008; Levitan, Maredia, and Hirschberg, 2018; Seeun Jung and Radu Vranceanu, "Experimental Evidence on Gender Differences in Lying Behaviour," *Revue Économique*, Vol. 68, No. 5, 2017, p. 859; and Jussi Palomäki, Jeff Yan, David Modic, and Michael Laakasuo, "To Bluff Like a Man or Fold Like a Girl? Gender Biased Deceptive Behavior in Online Poker," *PloS One*, Vol. 11, No. 7, 2016.

[2] Dou et al., 2017; and Levitan, Maredia, and Hirschberg, 2018.

[3] Jung and Vranceanu, 2017; and Palomäki et al., 2016.

men's lies are more self-oriented while women's lies are more other-oriented.[4] For instance, one experiment found that men are more likely than women to tell *selfish black lies* for individual economic gain and *Pareto white lies* for mutual gain,[5] whereas women are more likely than men to tell *altruistic white lies* that benefit others, despite incurring personal tradeoffs.[6] However, a more recent meta-analysis contradicts these results by showing that, while men are significantly more likely than women to tell black lies, they are also more likely than women to tell altruistic white lies. Results were inconclusive as to whether gender differences exist in the case of Pareto white lies.[7]

In terms of the words that men and women use in deceptive communication, Dou and colleagues (2017) analyzed a large dataset of text messages and found that, compared with men, women used more self-oriented words (e.g., *I, I'm*) and fewer other-oriented words (e.g., *you, your*) while lying across the board. Levitan and colleagues (2018) analyzed linguistic features in truthful and deceptive interview questions and found that the use of family-related words and articles (e.g., *the, a/an*) were indicative of deceptive speech among women.

Using the literature as a guide, we conducted an inductive analysis to determine whether the accuracy of our models differed significantly between men and women. Our findings show evidence of differing results based on the gender of the interviewee. That is, our models have a higher rate of accuracy for detecting deception among men interviewees than women interviewees in our sample.

Our Models Were More Accurate at Detecting Lies by Men Than Women

To determine whether our model's accuracy rates varied between men and women interviewees, we re-ran the models while taking gender into account. Because of limitations in our data collection, we knew the gender for only 99 out of 103 participants in our sample. Of the 50 men in our sample, 32 were assigned to lie while 18 were assigned to truth-telling conditions. Of the 49 women in this sub-sample, 24 were in the lying condition and 25 were in the truth-telling condition.

Table 5.1 shows performance metrics for each model both overall and broken down by gender. We note that the models displayed in this table were trained on different subsets of the data (i.e., subset of men, subset of women, or the entire sample of men and women

---

[4] DePaulo et al., 2011.

[5] *Selfish black lies* are defined as lies "involving acts that help the liar at the expense of another" (Erat and Gneezy, 2011, p. 723). *Pareto white lies* exist "when both sides earn more as a result of the lie." Selfish black lies are defined as lies "involving acts that help the liar at the expense of another" (Erat and Gneezy, 2011, p. 724).

[6] *Altruistic white lies* are ones that "can harm the liar but help the other person" (Erat and Gneezy, 2011, p. 724); and Erat and Gneezy, 2011.

[7] Valerio Capraro, "Gender Differences in Lying in Sender-Receiver Games: A Meta-Analysis," *Judgement and Decision Making*, Vol. 13, No. 4, July 2018.

together). Thus, systematic differences between model results from the subsets of men or women might differ from a model applied to a combination of these subsets.

Overall, we found that the word-counting model continued to provide the best combined accuracy for detecting lies—approximately 76 percent accuracy. However, when we disaggregated the models using gender, we found that our logistic regression classifier was significantly more accurate at detecting deception among men than women, most prominently in the word-count model: There was a 20.5-percentage point gap in the accuracy rate.

We also found that the word *I* was associated with detecting deception among men. That is, men were less likely to use *I* when lying and more likely to use it when telling the truth (Figure 5.1). This difference was statistically significant (a *p*-value of 0.0015 for a two-sample Kolmogorov-Smirnov test).

This finding partially helps explain why the word-count model is more efficient at detecting lies versus truths for men than for women. This feature is the most explanatory feature for men and therefore the best predictor for improving model accuracy for men (Figure 5.2).

TABLE 5.1
Classification Performance by Gender

Model	Subgroups	Accuracy (percentage)
Word count	Overall	75.6
	Men	76.0
	Women	55.5
Stance vectors	Overall	64.1
	Men	62.0
	Women	61.5
Metadata	Overall	62.7
	Men	64.0
	Women	66.5
Word count and metadata	Overall	74.5
	Men	80.0
	Women	67.0
Word count, metadata, and stance	Overall	62.2
	Men	58.0
	Women	57.5

NOTE: This table shows accuracy scores for the different models developed, broken down by gender of interviewees (men and women) for the interviews. The simple word-counting model combined with metadata is the most accurate. Stance appears to be a relatively poor predictor of deceptive speech in general.

FIGURE 5.1

Word Usage by Gender

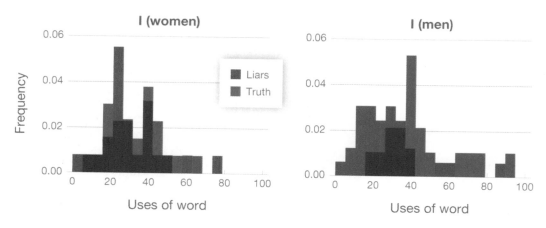

NOTE: This figure shows the relative number of uses of one word of interest (*I*) for men versus women, identified during the word-importance analysis shown in Figure 5.2. Men used the word *I* less often when they were lying and more often when they were telling the truth. This finding is statistically significant ($p < 0.01$; two-tailed). For women, however, there were no significant differences between liars and truth-tellers in the usage of this word.

Additionally, this finding indicates that the data themselves differ by gender, and as a result, the models are in some way picking up on this variation. Put differently, it is not the models themselves that are biased, but the models likely are picking up on certain speech pattern differences between men and women that then make it easier to detect when men are lying. These differences could be causing the models to produce unequal accuracy rates for men and women.

We also uncovered other notable speech pattern differences between men and women. For instance, women were significantly more likely than men to use the word *man*. The word is used generally in the context of the interview (i.e., the interviewees were asked to read an article or document about a man who leaked information and then respond to a question about what they had read). Although not indicative of deception, it might suggest that the gender of the subject during this task was more salient for women than it was for men and therefore might have differentially shaped their responses to interview questions.

FIGURE 5.2
Word Importance by Gender

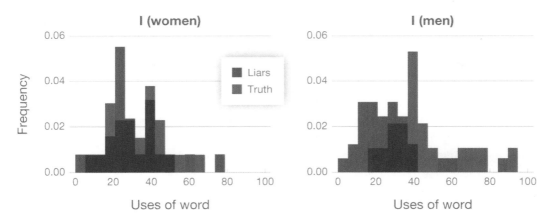

NOTE: This figure shows the relative number of uses of one word of interest (*I*) for men versus women, identified during the word-importance analysis shown in Figure 5.2. Men used the word *I* less often when they were lying and more often when they were telling the truth. This finding is statistically significant ($p < 0.01$; two-tailed). For women, however, there were no significant differences between liars and truth-tellers in the usage of this word.

Does the Interviewer's Gender Matter?

The aforementioned findings led us to ask the following question: What factors might account for the different speech patterns observed between men and women interviewees? The literature often discusses lying as an interactional process in which peoples' expectations about others shape interaction patterns.[8] In other words, when individuals lie, they likely are influenced by the people to whom they are lying and vice versa.[9] For instance, interviewees

---

[8] DePaulo et al., 1996.

[9] Hancock et al., 2008.

answering sensitive questions might feel more at ease with interviewers of their same gender and therefore might be more inclined to provide more-honest responses.[10]

In this study, all trained interviewers were men. Research has shown that men are more likely to attempt to deceive women than other men because of widely held stereotypes about the gullibility of women.[11] Therefore, it is possible that male interviewees in the study found it more challenging to lie to a male interviewer than if they had been paired with a female interviewer. It also is possible that the male interviewers spoke differently to interviewees depending on their gender (e.g., a more aggressive voice associated with men and a softer voice associated with women), causing variation in the responses as a result.[12] Another possibility is that the female interviewees self-monitored or spoke more carefully in some way because they were being interviewed by a man instead of a woman.[13] Further analysis is required to confirm the effects of interviewer gender in the data.

Implications and Key Considerations

These results provide suggestive evidence that gender differences in deceptive speech could cause the automated tools for detecting deception to produce biased results. These findings also reinforce that gender should be a variable in the collection and analysis of interview data when conducting deception-related research. The accuracy gap observed in the results was not negligible; future data-collection efforts should include both men and women interviewers to prevent the potential bias in interviewee responses.

From a national-security and human-capital perspective, there are several important implications to keep in mind. First, if men are conducting many of these background investigations, it might affect the results of these models. If government officials choose to use automated tools for detecting deception down the road, they should recognize both the advantages and limitations that come with using these tools. More specifically, government officials should recognize that the inappropriate use of automated tools could lead to inequitable acceptance and rejection rates among individuals who apply for security clearances. When it comes to recruiting the best talent in the federal government, this limitation could have meaningful implications in terms of both national security risks and diversity, equity, and inclusion efforts. Moreover, these findings imply that there could be implications beyond gender and that other ascribed characteristics or demographic traits (e.g., age, national origin, race) could affect deception detection results.

---

[10] Lipps and Lutz, 2017.

[11] Palomäki et al., 2016; and Jung and Vranceanu, 2017.

[12] Leongómez et al., 2017.

[13] Lipps and Lutz, 2017.

Finally, these findings reinforce the need to maintain human-in-the-loop (HITL) calibration to mitigate the risks that come with using ML approaches in the realm of national security. These risks include factors, such as algorithmic bias and false positives and false negatives. However, HITL is not necessarily a total solution because humans have their own biases as well. Therefore, it is important to maintain an adversarial process with a system of checks and balances in which neither human nor machine trust the other completely as a means of risk prevention.

CHAPTER SIX

Limitations, Conclusions, and Recommendations

This report presents exploratory results on the use of automated tools to detect attempts at deception during a study that simulates part of a security clearance investigation interview. Typically, applicants for a clearance will complete a Standard Form (SF) 86, and then review their answers with a trained interviewer. We propose that how interviewees answer these questions could be a useful signal to detect when they are trying to be deceptive. Specifically, clearance interviews represent social interactions in which, typically, two people (interviewer and interviewee) express and interpret both verbal and nonverbal cues between each other. We proposed that ML models have the capacity to detect some of these signals.

This report describes an exploratory study in which participants read the same document (randomly assigned to participants so it was presented in two different formats), and then were instructed to be truthful or deceptive in interviews (conducted using MCI) about what they had just read. A subcontractor audio recorded these interviews and collected text-based chat logs, RAND then transcribed the audio files, and then RAND developed several ML models that detected deception versus truthfulness using features of interviewee answers from both the chat logs and audio files. Results from this exploratory analysis identified several linguistic patterns on which our models could be trained to predict attempts at deceitfulness.

Limitations

This research has several limitations. First, it is exploratory research that tried several different modeling techniques on a relatively small sample of 103 participants associated with the University of New Haven. Thus, our sample is neither representative of all adults in the country nor of those applying for a security clearance. Furthermore, our subcontractor had problems with data collection that resulted in more participants being assigned to the lying condition than the truth-telling condition. Thus, a larger study that ensures correct randomization is required to ensure accurate comparisons.

Second, there is likely variation in interviewer training and approach. Our subcontractor recruited former agents from the FBI to be interviewers. Although they were randomly assigned to participants, their interviewing style likely affected the interviews and the data. Chapter Five reviewed a subsample of these data to analyze the potential for bias based on the gender of participants because all four interviewers were men.

Third, there were several interruptions during data collection. Our subcontractor began to collect data before the coronavirus pandemic began, there were some interruptions during this pandemic, and the study continued throughout the pandemic. Furthermore, our subcontractor had several staff members help collect data over the past several years, and inconsistencies among interviewers and data collectors could have affected the collection of these data.

Fourth, the content of the interviews might have played a role in model performance. Each interview was very similar, asking the same kinds of questions about the same topics. It is unclear whether a security clearance interview, which covers many distinct topics, would see similar performance for our models. Although many of the features that we found to be indicative of deception were consistent with the literature on lying in other circumstances (the difference between autobiographical and you/attention stance, for instance), others might not translate well to other domains. In particular, the observed pattern in the use of the word *truth* (which referred to the video that participants watched about the importance of telling the truth) is unlikely to be seen in actual clearance interviews. On the other hand, the fact that the word *truth* was seen as significant might suggest the salience of this topic to interviewees' psychology of deception and could perhaps be useful going forward.

Fifth, our study design asked participants to lie about an event that likely had little meaning to them: reading a short news story (or memo) about someone else who leaked classified information. Thus, participants who are better at telling lies in general (e.g., being successful at deception in the past) might find it easier to lie in low-stakes situations, such as in this study. Furthermore, we cannot assess whether actual interviewees might be more or less likely to lie during interviews conducted in-person or via VTC, independent of the patterns in linguistic signals that we detected. These limitations demonstrate a need for further research to refine our models using larger samples that are ideally representative of those who apply for security clearances.

Finally, we considered potential gender bias in Chapter Five, but other forms of bias likely exist in our data and analysis. One possible source could be dialect: The race, age, economic condition, and geographical origin of individuals in our sample (participants associated with a college campus in the northeastern United States) means that the population we studied is unlikely to be nationally representative. There are a multitude of dialects in American English, and each likely has subtle implications for how its users lie. We did not have a sufficiently large or detailed database in this study to examine the effects of dialect on our models, but implementation of an automated deception-detection system in the security clearance process should consider this possible source of bias as well.

Conclusions and Recommendations

We reached four conclusions based on this exploratory research, which led us to make the six recommendations that we describe in this section.

First, linguistic signals exist when people attempt to be deceptive, and ML tools have the capacity to detect some of these signals for interviews conducted via VTC and text-based chat. For both interview modes, models that used word counts were some of the most accurate at predicting who was trying to be deceptive versus those who were truthful. Our word-vector model had an accuracy rate of 75.6 percent for VTC interviews and 65.6 percent for chat logs. Combining different factors (i.e., word counts, stance, and metadata) did not produce accuracy rates that were higher than simply using word count. This leads to our first recommendation.

Recommendation 1: The federal government should test ML modeling of interview data that uses word counts to identify attempts at deception.

Second, we found similar differences in accuracy rates for detecting deception when interviews were conducted over VTC and text-based chat. Our subcontractor randomly assigned participants to conduct the same interview with the same questions over VTC or chat in random order. For both interview modes, models using word counts were some of the most accurate predictors of who was assigned to lie or tell the truth. These results suggest that chat might be a useful means to conduct security clearance background investigation interviews, but we caution that further pilot testing is required before the federal government scales chat-based interviews. Furthermore, we found evidence that models using both the chat logs and VTC transcriptions together produced a 71.8 percent accuracy rate of predicting liars from truth-tellers. This leads to our next two recommendations.

Recommendation 2: The federal government should test alternatives to the in-person security clearance interview method—including VTC and chat-based modes—for certain cases.

Recommendation 3: The federal government should test the use of asynchronous interviews via text-based chat to augment existing interview techniques. The data from an in-person (or in-person virtual) interview *and* chat could help investigators identify topics of concern that merit further investigation.

Third, this exploratory study developed a workflow for transcribing and analyzing interview data that was audio recorded. The workflow involved using AWS Transcribe, which used ML to automatically transcribe each interview recording; having a member of the research team listen to this audio and correct each transcript; then using our models to analyze these revised transcripts. Although the general substance of the AWS Transcribe transcriptions were correct, we found several errors when reviewing the output. None of these automated transcriptions were free of errors, and they often missed subtle features of informal speech (e.g., "uhh" or laughter). Although additional research is required before the government can begin to use ML methods to transcribe and analyze interview data, we note that there is promise in piloting these evolving methods to augment existing processes during the security clearance process. This leads to our next recommendation.

Recommendation 4: The federal government should use ML tools to *augment* existing investigation processes by conducting additional analysis on pilot data, but the government should not replace existing techniques with these tools until they are sufficiently validated.

Fourth, we found evidence of gender bias in our exploratory models. For example, our models that used word counts produced the highest accuracy rates for all participants in our sample (75.6 percent). However, the accuracy rate was 76.0 percent for men in our sample and 55.5 percent for women. Across most of our models, we found evidence that they were more accurate for men than women. There are several potential explanations for this discrepancy, including the ways in which men and women might answer the same questions during these interviews and the fact that all of the interviewers were men. Although our sample size was not large enough to analyze model accuracy by other demographic characteristics (e.g., participant's age or race), and this analysis was done on a subset of data in which the gender of participants was collected, we suspect model accuracy might vary by other features as well. Thus, we make the following recommendations.

Recommendation 5: The federal government should validate any ML models that it uses for security clearance investigations to limit the bias in accuracy rates on the basis of the ascribed characteristics (e.g., race, gender, age) of interviewees.

Recommendation 6: The government should have HITL to continuously calibrate any ML models used to detect deception during the security clearance investigation process.

Potential for Next Steps

We conclude this report by describing two ways in which federal government could use these results in the future. First, the government should pilot test these ML models on a subset of low-risk applicants and, after rigorous review, scale these models to larger numbers of applicants. Furthermore, the government should use such models as one tool in its toolbox for detecting deception during the security clearance investigation process. Put simply, these models should augment existing processes, not replace them entirely.

Second, the government could use these models to augment its existing practices in several ways. For example, if there are concerns that a security clearance applicant is lying about a particular topic (e.g., foreign contacts) on their SF-86, investigators could use a standard script for a follow-up interview, record this interview, transcribe it, and then use models like those described in this report to detect the potential for deception. Our word-count model might give insights into whether the government should spend additional resources to investigate an applicant's foreign contracts—or our deep learning contradiction model (proof of concept, described in Appendix D) might identify specific points in the applicant's story that merit further investigative resources. Although our exploratory models cannot confirm with certainty that someone is lying, these models could have the capacity—after additional refining and pilot testing—to help the government prioritize resources when it is looking for lies.

Modified Cognitive Interviewing

Cognitive interviewing (CI) is one type of method that has been used by researchers and law enforcement to detect deceptive speech. Developed in the 1980s by Geiselman and Fisher,[1] CI is based on two principles of memory retrieval and is carried out using four mnemonic techniques. The two memory retrieval principles stipulate that: (1) mentally recreating the context of an event facilitates memory,[2] and (2) more than one mental pathway or cue can lead to the same memory.[3] The four mnemonics techniques used for memory retrieval in CI are (1) reporting everything, (2) reinstating mental context, (3) changing temporal order, and (4) changing perspective.

Although CI initially was able to improve interviewees' memory retrieval better than standard police interviews, some research later showed that many police officers found CI too burdensome and ill-adapted to the settings in which they conducted their interviews. As a result, officers often neglected to fully implement CI. In recent years, researchers have sought to improve on the CI method by modifying it and applying its principles to an enhanced CI, sometimes referred to as *MCI*.

To enhance the performance of CI in applied settings, researchers modified the CI in two important ways: (1) by meaningfully shortening the CI question protocol and making its prompts less formal, and (2) by adding evidence-based communication techniques. These techniques include rapport-building with the interviewee, transferring control of the interview to the interviewee by not interrupting their recall of events, and structuring the inter-

---

[1] R. E. Geiselman, R. P. Fisher, I. Firstenberg, L. A. Hutton, S. J. Sullivan, I. V. Avetissain, and A. L. Prosk, "Enhancement of Eyewitness Memory—An Empirical Evaluation of the Cognitive Interview," *Journal of Police Science and Administration*, Vol. 12, No. 1, March 1984.

[2] Endel Tulving and Donald M. Thomson, "Encoding Specificity and Retrieval Processes in Episodic Memory," *Psychological Review*, Vol. 80, No. 5, 1973.

[3] Endel Tulving, "Cue-Dependent Forgetting," *American Scientist*, Vol. 62, No. 1, January–February 1974; and Gordon H. Bower, "A Multicomponent Theory of the Memory Trace," in K. W. Spence and J. T. Spence, eds., *The Psychology of Learning and Motivation: Advances in Research and Theory*, Vol.1, New York: Academic Press, 1967.

view in a way that is compatible with the ordering of events in the interviewee's mental record.[4] These methodological enhancements resulted in MCI.

By asking interviewees to explicitly recount everything they can remember—regardless of whether they believe it is important—MCI increases an interviewer's exposure to information and makes detecting deception more likely. Interviewers ask interviewees to recount their memories using four prompts (i.e., visual, auditory, personal feelings, and temporal reversal). Because the interviewee controls much of the direction and content of the interview, the interviewer has many opportunities to pick up on speech-related signals of deception. The structure of the interview also allows the interviewer to further probe suspect information and better ascertain interviewee deception. As a result, interviewees are more likely to experience higher levels of cognitive load or mental taxation, which can more-easily allow interviewers to elicit speech content differences among deceptive versus truthful actors. To date, MCI has been shown to yield deception detection rates (i.e., 80 to 85 percent) that are significantly higher than those achieved by chance (i.e., 50 percent) or by human judgements (i.e., 54 to 56 percent).[5] With respect to the interview data analyzed in this study, all interviewers used MCI during the interviews.[6]

---

[4] Cindy Colomb and Magali Ginet, "The Cognitive Interview for Use with Adults: An Empirical Test of an Alternative Mnemonic and of a Partial Protocol," *Applied Cognitive Psychology*, Vol. 26, No 1, January–February 2012.

[5] Morgan et al., 2013.

[6] In addition to this study, our subcontractor collected data on a separate study that compared MCI with a traditional interview technique. Results from analysis of these data may be published in other non-RAND outlets on a future date.

Study Materials

This appendix displays some of the materials used in the study that was administered by our subcontractor, the University of New Haven. Figure B.1 displays the demographic questionnaire that participants completed before starting this study. Figure B.2 shows the fictitious news story about Edward Snowden leaking classified information. Figure B.3 displays a fictitious sensitive memo with markings that uses the same text from the news story. Figure B.4

FIGURE B.1
Demographic Questionnaire for Study Participants

Demographics Questionnaire

Please complete the following:

Subject #: _____

Age: _____

Gender *(Please circle one)*: M / F / Other

What is your Race? _____

What is your Ethnicity? _____

Marital Status: _____

Highest Level of Education: _____

Household Income: _____

displays a screenshot of a video by a retired FBI special agent instructing participants to tell the truth before they began their interviews. Finally, the last part of this appendix displays the study instructions that staff from the University of New Haven followed.

FIGURE B.2
Fictitious News Story About the Leaking of Classified Information

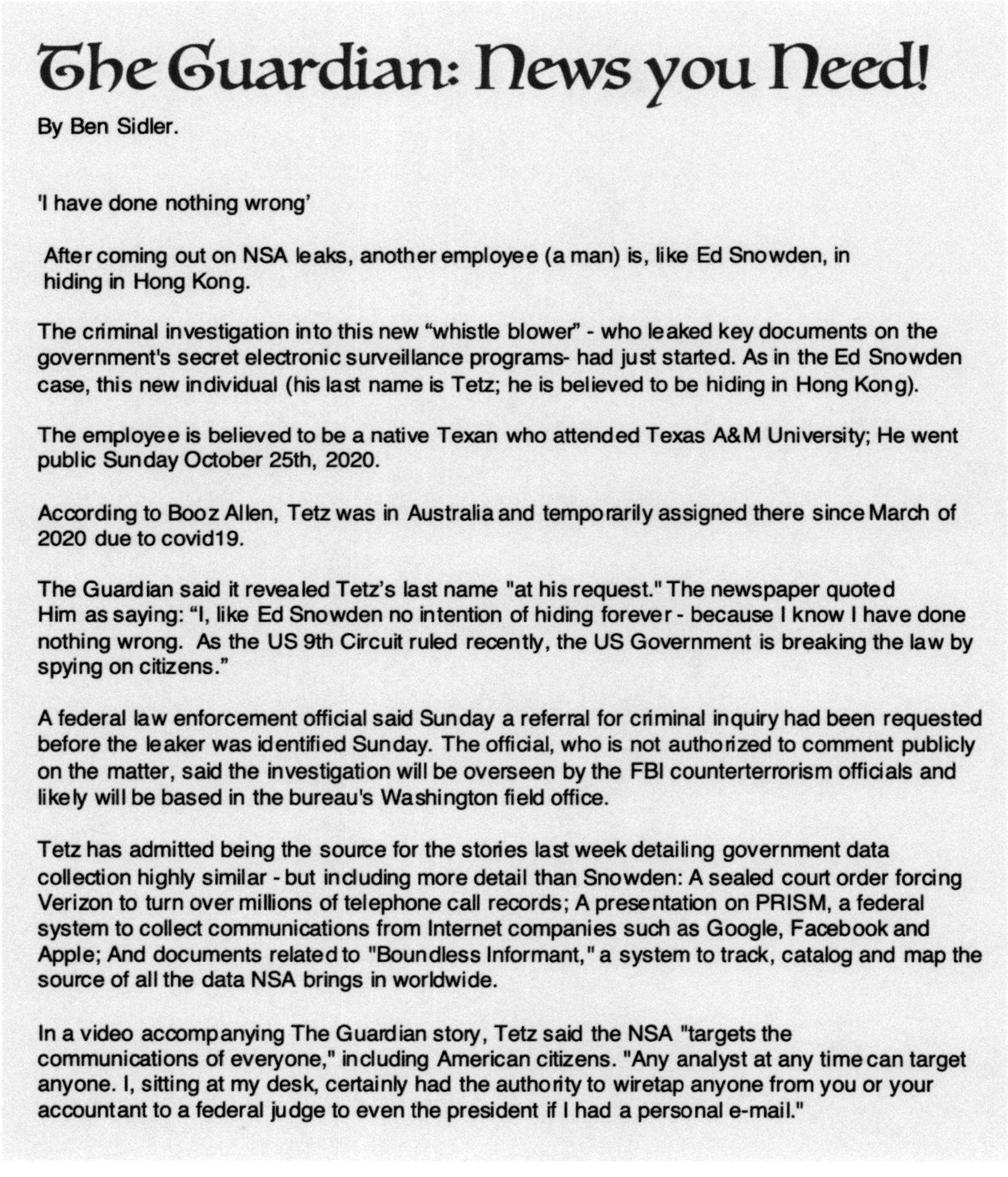

The Guardian: News you Need!

By Ben Sidler.

'I have done nothing wrong'

After coming out on NSA leaks, another employee (a man) is, like Ed Snowden, in hiding in Hong Kong.

The criminal investigation into this new "whistle blower" - who leaked key documents on the government's secret electronic surveillance programs- had just started. As in the Ed Snowden case, this new individual (his last name is Tetz; he is believed to be hiding in Hong Kong).

The employee is believed to be a native Texan who attended Texas A&M University; He went public Sunday October 25th, 2020.

According to Booz Allen, Tetz was in Australia and temporarily assigned there since March of 2020 due to covid19.

The Guardian said it revealed Tetz's last name "at his request." The newspaper quoted Him as saying: "I, like Ed Snowden no intention of hiding forever - because I know I have done nothing wrong. As the US 9th Circuit ruled recently, the US Government is breaking the law by spying on citizens."

A federal law enforcement official said Sunday a referral for criminal inquiry had been requested before the leaker was identified Sunday. The official, who is not authorized to comment publicly on the matter, said the investigation will be overseen by the FBI counterterrorism officials and likely will be based in the bureau's Washington field office.

Tetz has admitted being the source for the stories last week detailing government data collection highly similar - but including more detail than Snowden: A sealed court order forcing Verizon to turn over millions of telephone call records; A presentation on PRISM, a federal system to collect communications from Internet companies such as Google, Facebook and Apple; And documents related to "Boundless Informant," a system to track, catalog and map the source of all the data NSA brings in worldwide.

In a video accompanying The Guardian story, Tetz said the NSA "targets the communications of everyone," including American citizens. "Any analyst at any time can target anyone. I, sitting at my desk, certainly had the authority to wiretap anyone from you or your accountant to a federal judge to even the president if I had a personal e-mail."

FIGURE B.3

Fictitious Sensitive Memorandum About the Leaking of Classified Information

To: Ambassador Culvahouse
Re: NSA Whistleblower
Level of Importance: HIGH
Classification: SENSITIVE

(S) After coming out on NSA leaks, another employee (a man) is, like Ed Snowden, in hiding in Hong Kong. The criminal investigation into this new "whistle blower" - who leaked key documents on the government's secret electronic surveillance programs- had just started. As in the Ed Snowden case, this new individual (who remains anonymous) has been an employee of the NSA for over 15 years. This individual is 45 and has claimed responsibility for the current release of classified information.

(S) Now, this NSA employee (Ron Tetz) is hiding in a Hong Kong hotel room (the downtown Marriott); He, like Snowden, is seeking to avoid criminal prosecution. He has told The Guardian, the British newspaper to whom Snowden also leaked documents, that he "is not afraid."

(S) Tetz is a native Texan who attended Texas A&M University; He holds a TS-SCI clearance and was stationed at the NSA post in Australia. He went public on Sunday October 25th, 2020. They indicated they are still assessing the damage - but fear it is worse than the damage caused by Ed Snowden.

(U) The National Security Agency press release indicated the matter regarding the whistleblower has been referred to the Department of Justice.

(U) Any person who has a security clearance knows that he or she has an obligation to protect classified information and abide by the law," said Shawn Turner, a spokesman for the Office of the Director of National Intelligence.

(U) The employee is a former NSA technical professional who worked initially for the CIA, then moved to the NSA. He was with the NSA, he was a contractor and paid by Booz Allen. Booz Allen has confirmed Sunday that Tetz, like Snowden, had been an employee. Unlike Snowden, Tetz had worked for Booz Allen for 10 years. According to Booz Allen, Tetz left the NSA's post in Australia 3 weeks ago and took classified documents with him.

(U) The Guardian said it revealed Tetz's identity "at his request." The newspaper quoted Him as saying: "I, like Ed Snowden, have no intention of hiding forever - because I know I have done nothing wrong. As the US 9th Circuit ruled recently, the US Government is breaking the law by spying on citizens."

(U) A federal law enforcement official said Sunday a referral for criminal inquiry had been requested before the leaker was identified Sunday. The official, who is not authorized to comment publicly on the matter, said the investigation will be overseen by the FBI counterterrorism officials and likely will be based in the bureau's Washington field office.

(U) Tetz has admitted being the source for the stories last week detailing government data collection highly similar - but including more detail than Snowden: A sealed court order forcing Verizon to turn over millions of telephone call records; A presentation on PRISM, a federal system to collect communications from Internet companies such as Google, Facebook and Apple; And documents related to "Boundless Informant," a system to track, catalog and map the source of all the data NSA brings in worldwide.

(U) In a video accompanying The Guardian story, Tetz said the NSA "targets the communications of everyone," including American citizens. "Any analyst at any time can target anyone. I, sitting at my desk, certainly had the authority to wiretap anyone from you or your accountant to a federal judge to even the president if I had a personal e-mail."

FIGURE B.4

Introductory Video Before Interview by Former FBI Special Agent

Demographics Questionnaire

Please complete the following:

Subject #: _____

Age: _____

Gender *(Please circle one):* M / F / Other

What is your Race? _____

What follows is a description of the order of operations that our subcontractor used for data collection and was provided to RAND:

1. (RECORD) Review consent form with subject. Ask them to read the final statement out loud instead of signing it. Also ask the subject to send an email to the lab saying they have read the consent for and wish to participate in the study. Remind the subject they will be paid over Venmo. Instruct the subject that you will ask them the question printed at the bottom of their consent form, and that they are going to lie about it. (STOP RECORDING)
2. Have the subject fill out the demographics form. Remind them that this is not part of their task.
3. (RECORD) Show subject deceptive pre-video. Remind them that this is when their task starts when the video does.
4. Inform the participant that they will have approximately 10 minutes to read through a reading and memorize as many details as they can. They may read at their own pace.
5. Inform them that their task is over.
6. **[If the participant is in the lying condition, say the following in italics:]** *"Tell them they have been assigned to the partial lie condition for their interviews. Instruct the subject that they are not to reveal that they read a read a sensitive document and should claim they read a news article instead. They also must not mention any information found in the RED text, but should otherwise be honest about what they did read that was not in RED."*

7. They should only speak about the task when interviewed that the interviewers know about the other items they have already done for the study (the Self-Assessment Manikin [SAM]), Demographics form, the other interview). The interview is only focused on their task. Tell them their first interview will be with an AI system that is generating the questions. They will type their answers. After completing this interview, they will have a second interview that will be with a live person doing the interviewing. (STOP RECORDING)

8. Subject fills out SAM mannequin

9. (RECORD) Inform the subject that they will now be asked questions generated from an AI [artificial intelligence] and that the questions will appear in the chat and that they should respond in the chat as well. Copy and paste the following questions one by one into the chat. Copy and paste their answers into a separate document.

   a. Think about your experience from the time you started the task until you finished. Focus on what you remember seeing and doing for both tasks. Say everything you remember in as much detail as possible. Be as detailed as you can be. Don't leave anything out—even if you think it is trivial or insignificant. IF you remember it mention it.

   b. From the beginning of the task. State your visual memory about your experience —everything you remember seeing with your eyes from the time it started until the time it ended.

   c. Go back to the beginning of the task. I'm interested in your auditory memory about your experience—everything you remember hearing with your ears from the time it started until the time it ended.

   d. What was the experience like for you personally? What feelings did you have during the task?

   e. Start with the last thing that is in your memory while completing this task and state the last thing you remember, then state what happened right before that, and before that, and so on walking your way backwards through your memory; all the way back to the beginning. Almost like a movie playing in reverse.

   f. Do you think you left anything out or made any mistakes in what you've stated about your experience performing the task with the researchers today? (STOP RECORDING)

10. (RECORD) Bring the interviewer on to the call. Record the interview. (STOP RECORDING)

11. Have the subject fill out the SAM again.

12. (RECORD) Ask the subject their unique question. (STOP RECORDING)

13. Get subject's Venmo information. Thank them for participating.

Example Output from Amazon Web Services Transcribe

Below is a sample output for a partial transcript that AWS Transcribe created for a participant.

Transcription of ND8918_143498

Transcription using AWS Transcribe automatic speech recognition and the 'tscribe' python package. Document produced on Thursday, July 1, 2021, at 15:30:34.

TABLE C.1
Transcription Using AWS Transcribe

Confidence	Count	Percentage
98–100 percent	965	79.75
90–97 percent	57	4.71
80–89 percent	14	1.16
70–79 percent	16	1.32
60–69 percent	12	0.99
50–59 percent	12	0.99
40–49 percent	3	0.25
30–39 percent	1	0.08
20–29 percent	0	0.0
10–19 percent	0	0.0
0–9 percent	0	0.0

FIGURE C.1

Screenshot of Plot from AWS Transcribe

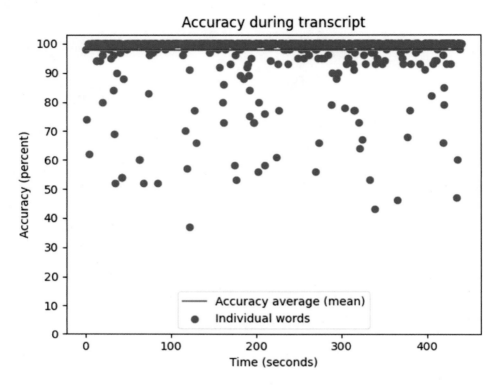

SOURCE: Amazon Web Services, "Amazon Transcribe," webpage, undated.
NOTE: The y-axis represents estimates of accuracy for a word in the transcript. The x-axis represents the time per second in the audio recording being transcribed. The blue dots are individual words in the transcription. The red line is the average accuracy for all words transcribed.

TABLE C.2
Partial Transcript of Interview with Participant

Time	Speaker	Content
00:00:00	spk_0	also?
00:00:01	spk_0	Yeah.
00:00:03	spk_0	Hi, my name's Russ. Thanks for participating in this research project.
00:00:10	spk_0	Before we begin, I'd like to let you know that if any time you feel like you need to take a break, just let me know. You should take about 30 minutes. It's important that you remember whatever instructions were provided to you by the research team. But please don't tell me what those instructions were.
00:00:27	spk_0	If you're unclear about the instructions that were provided to, you can step outside and speak of the member of the research team to double check so that you're clear and for the purposes of this interview you've already filled out security clearance questionnaire. SF 86. All right,

NOTE: Gray text has less than 98 percent confidence.

Proof of Concept: Deep Learning Contradiction Model

This appendix presents a proof of concept for applying a deep learning model to our data for the purpose of detecting contradictions in the responses of interviewees. Consider an interviewee who, at first, denies having any financial difficulties when asked directly, but later mentions in passing that they are burdened by debt. Such an admission might be overlooked during a conversation, but thorough analysis of the transcript would show that the interviewee had not been fully truthful throughout the interview. We automate this process to quickly flag these kinds of subtle contradictions. While there is a need for relevant training data, this proof of concept showcases a potential use case for these models.

Methods

Identifying contradictions in text is difficult to do with the kinds of simple models that we described previously. Instead, we rely on general-purpose deep neural networks, based on the Transformer architecture, that have been trained to perform a variety of NLP tasks. Transformer networks consider the words in a sequence of text and the order and context in which they appear. This allows the models to be used for language translation, text summarization, question-answering, and even text synthesis. Research has also been done to train Transformer models for sentence entailment tasks (determining whether one sentence implies the other). We hoped that sentence entailment would be similar enough to our problem of contradicting statements that a model trained for one task would be able to transfer its capabilities to the other.

Transformer models rely on a mechanism called *attention* to understand the context of each word they are analyzing. The details of the attention mechanism are technical, but, for example, it allows the neural network to identify that the word *it* in the sentence "The person read the newspaper and then dropped it in the recycling bin" refers to the newspaper, not the recycling bin. By studying the detailed attention data of our trained model, we hoped to be able to pinpoint the exact words in a sentence that led to a contradiction. In short, building a contradiction detector could lead to a system that quickly identifies conflicting statements and then highlights where they appear to conflict, allowing a human to make a faster and more informed decision about the interviewee, potentially in real time.

We used the HuggingFace TransformerForSequenceClassification module,[1] training it on sentences taken from the Stanford Natural Language Inference (SNLI) dataset.[2] SNLI data are composed of pairs of short sentences and one of three labels: entailment, neutral, or contradiction. Because a label of entailment or contradiction should be independent of which statement comes first, we scrambled the input pairs so that the second statement came first for 50 percent of the time. We trained the network over three epochs and achieved an accuracy score of 87 percent—below the world record for this dataset but certainly adequate for a proof-of-concept model. We then applied our model to the pairs of responses in each of our interview transcripts, producing a matrix that indicates which combinations of sentences are in agreement or conflict.

Results

We concluded that the unique types of questions asked in the interviews from our limited sample of data and the contents of our training data were poorly suited to the task of contradiction detection. We had limited questions that were asked multiple times in the interviews, so most pairs of answers for both liars and truth-tellers are neutral. In future work, researchers could consider tailoring the interview style to the contradiction scheme described in this report—asking the same question in multiple ways, for instance. Our training dataset also was not well-suited for our task: Interview responses are in the first person, while the training dataset sentences are written in the third person; the pairs of training sentences in the SNLI dataset were written as a scene and then a description of the scene, not as multiple viewpoints of the same scene; and the training sentences are fairly short (a single sentence) compared with the interview responses (often several sentences). Again, these technical limitations could be lessened in future work if a large dataset of natural language responses to questions could be curated.

We did not apply our contradiction model to the chat data because each individual chat response generally was too long to be modeled well by our Transformer model. This is because only a handful of questions were asked in each chat-based interview, so the responses were necessarily longer than they were in the VTC interviews. Interpreting the attention results also would be extremely challenging when comparing two long responses. Incorporating the contradiction model into a chat-based system would require more questions with shorter responses.

Nevertheless, we did find some promising evidence that our approach could yield useful results if some of these data issues are mitigated. Figure D.1 displays one example of a contradiction matrix for Respondent #363. The red squares represent potential contradictions

---

[1] For more details, see Samuel R. Bowman, Gabor Angeli, Christopher Potts, and Christopher D. Manning, "A Large Annotated Corpus for Learning Natural Language Inference," *arXiv*, 1508.05326, August 21, 2015.

[2] Ashish Vaswani, Noam Shazeer, Niki Parmar, Jakob Uszkoreit, Llion Jones, Aidan N. Gomez, Łukasz Kaiser, and Illia Polosukhin, "Attention Is All You Need," *Advances in Neural Information Processing Systems*, Vol. 30, 2017.

between statements, the green squares are entailments, and blue squares are neutral in that they neither contradict nor imply one another. The vertical and horizontal bands of red indicate a set of statements that conflict with many other responses; in this case, the conflicting statements all pertain to remembering or misremembering certain facts. In particular, one answer (Statement 7) contains the phrase "my memory is definitely a little hazy," which is in contradiction with "I remember hearing" in Statement 4.

In Figure D.2, we display the internal model weights for these two different statements, both made by Respondent #363. The y-axis represents words for Statement 7: "Uh, my memory is definitely a little hazy." The x-axis represents words for Statement 4: "I remember hearing, like, why they, like why people would tell, like tell a lie."

FIGURE D.1

Contradiction and Entailment Matrix for Respondent #363 (Liar)

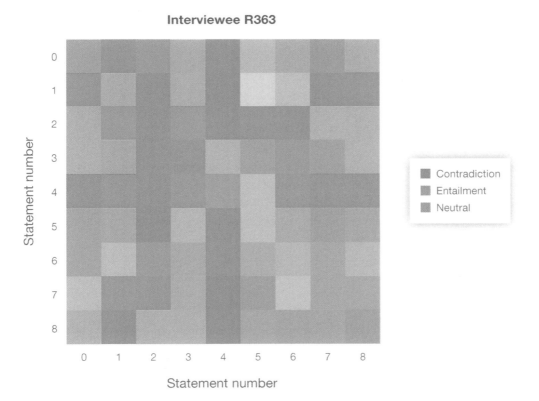

NOTE: This figure shows a matrix of interviewee responses that are color-coded by whether our deep learning model judged the statements associated with the row and column indices to be in contradiction, entail one another, or be neutral. In this case, most statements were either in agreement or neutral with one another, but a few statements—particularly those related to having an unreliable memory—appeared to be contradictory with many other statements.

FIGURE D.2

Attention Scores for Statements for Respondent #363 (Liar)

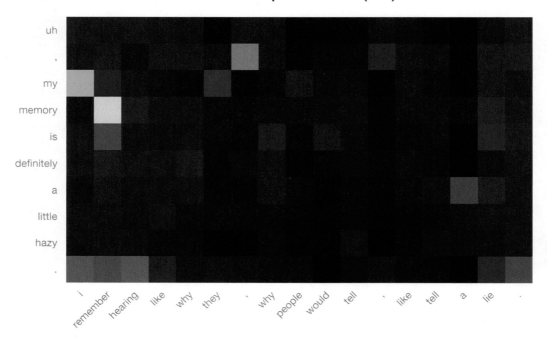

NOTE: This figure demonstrates an attention score matrix (the word combinations that matter most to the model decisionmaking) for a sample from one of the pairs of sentences that was judged to be contradictory. Yellow colors indicate high attention, while blue and purple indicate low attention.

The yellow box at the intersection of the words *memory* and *remember* shows that our deep learning model assigned larger weights to these words compared with other word pairs, which makes sense because they refer to similar topics. Theoretically, a more advanced system could use this kind of data to pinpoint the exact words that indicate contradiction within a longer answer. Still, the model does not seem to pick up on *hazy* as being a key part of the contradiction, as one might expect.

Such mixed results as these were present in other interviews. In Figures D.3 and D.4, we show analogous entailment matrices for two other interviews, demonstrating that the response of the neural network varied widely depending on the interviewee. In the case of Figure D.3, most of the responses from Respondent #351 appear to agree with one another, which is not surprising because this respondent was assigned to a truth-telling condition.

FIGURE D.3

Contradiction and Entailment Matrix for Respondent #351 (Truth-Teller)

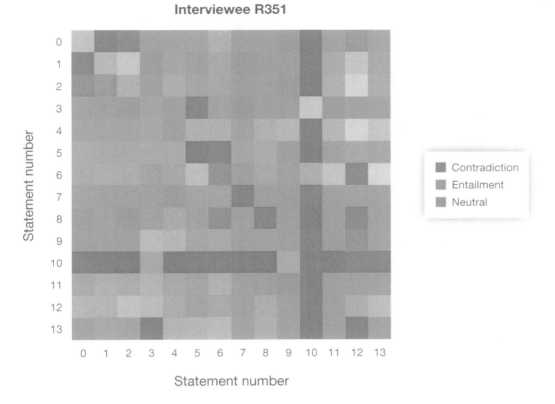

NOTE: This figure shows a matrix of interviewee responses for interviewee Respondent #351, who was a truth-teller. Most statements appeared to be in agreement with one another, with a few exceptions.

As we did earlier, we selected two of these statements that exemplified this agreement and examined their attention scores. On the *y*-axis of Figure D.4, we have the statement: "I was, [PAUSE] found it enjoyable." The *x*-axis shows the statement: "All right. It was, [PAUSE] it was [PAUSE] quite enjoyable." The word pair (*enjoyable* and *enjoyable*, in this case) indicate which words the model was using to inform its decision that the statements were in agreement with one another.

FIGURE D.4

Attention Scores for Statements for Respondent #351 (Truth-Teller)

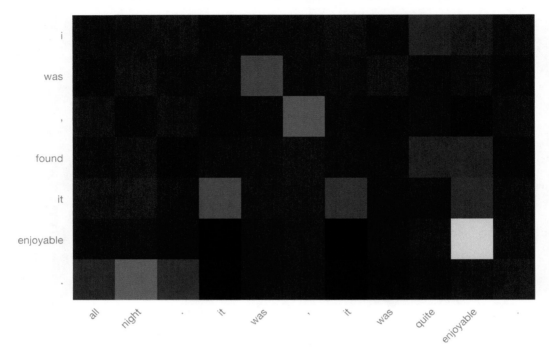

NOTE: Attention scores for two responses of respondent #351 (statements 9 and 13 in Figure 4.9). Strong entailment is seen for this pair of statements, which makes sense because they both express that the process was enjoyable. Unsurprisingly, the attention scores indicate that the presence of the word "enjoyable" in both sentences appears to be the driving factor behind this entailment classification.

However, not all of our results were so easy to interpret. In Figure D.5 (Respondent #262), most of the statement pairs were determined to be neutral, with little indication of any statements that were in obvious contradiction with many others.

In summary, we tested a separate deep learning model that takes all of the statements made by an interviewee, compares these statements with one another for this person, and then classifies statements that appear to be contradictory. We tested this model on a participant who first claimed her or his memory was "a little hazy" but later said they did in fact "remember seeing" something. Our deep learning model was capable of identifying this potential contradiction.

FIGURE D.5

Contradiction and Entailment Matrix for Respondent #262 (Liar)

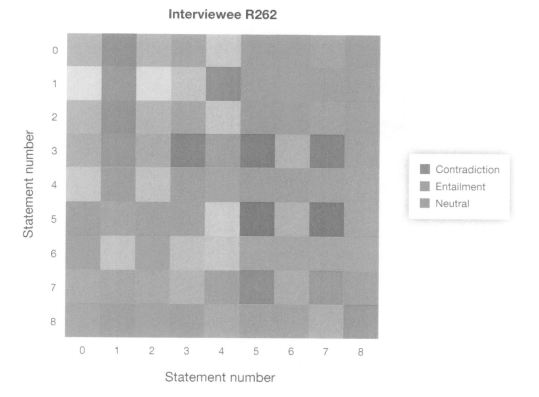

NOTE: This figure shows a matrix of respondent #262's responses. Here, we see that a lying interviewee may have a perfectly neutral matrix of responses without a clear pattern of contradictions. We take this as a sign that more research is needed to build more-successful models and apply them in useful ways.

The attention scores also show the difficulty of this approach. Figure D.6 displays the two statements on the *y*- and *x*-axes, respectively: "It was a little bit difficult. I may have just kind of jumbled up the information," and "So the feelings were more curiosity. I was interested to learn more about the topic." In this case, it is difficult to see how the model interprets these

FIGURE D.6

Attention Scores for Statements for Respondent #262 (Liar)

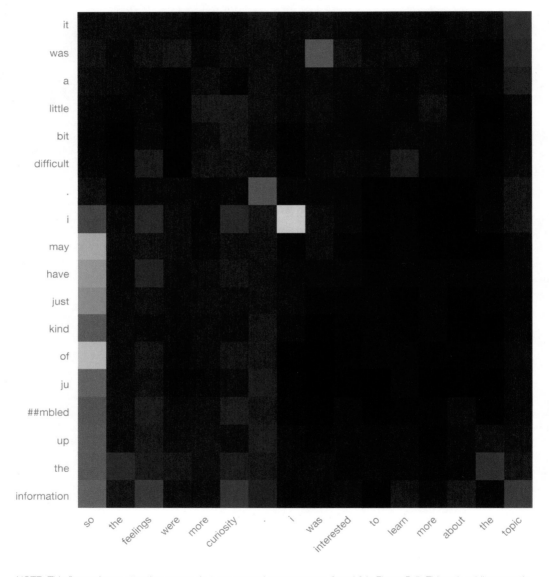

NOTE: This figure shows attention scores that correspond to statements 6 and 8 in Figure D.5. This pair, while correctly categorized as neutral (a statement that the task was difficult, and another that the task was interesting), indicates some of the difficulty in directly interpreting the attention scores of these kinds of deep learning models.

two sentences. We hope that future research can help build more-interpretable neural networks that would lead to a more robust contradiction model.

Although we did not attempt to measure the overall performance of our contradiction model in this proof of concept, we envision a future system that can combine the best aspects of our different models into a single HITL detector. The word choices used by interviewees could inform a top-level decision (a likelihood of lying), which could be further investigated with a detailed, answer-by-answer approach using a contradiction model. It might also be the case that certain metadata markers only appear when applied at an answer-by-answer level, but these markers are washed out when compiled across an entire interview; we leave this and other lines of inquiry for future work.

To operationalize this approach more successfully, we estimate that a dataset of tens of thousands of examples of natural language contradictions would need to be curated. These examples could be taken from trial transcripts or synthetically generated as a bespoke solution for this type of model. In addition, the interview process would need to be updated to ask about the same kinds of topics multiple times, preferably at different points throughout the interview. That structure would give the model more chances to catch contradictions while minimizing the probability that the interviewee could maintain their lie. Expert opinion on how to structure interviews to maximize the chance of arriving at a contradiction could also be elicited.

Potential Next Steps

We developed an exploratory deep learning model that identifies all the statements made by interviewees during our study, compares the content of these statements with one another, and then classifies statements that appear to be contradictory. We found evidence that this technique could work for some interviewees, but there are several limitations that the federal government would need to address. First, our training data for this deep learning model was not based on security clearance interviews. Second, interviewers asked different types of questions about the same topic throughout the interview (instead of asking the same question in different ways), making it difficult for this model to detect contradictions. However, we note that this type of modeling—when validated with additional pilot data—could be added to a toolbox when deciding whether to further investigate potential red flags that emerge during a security clearance background investigation. Thus, this type of model is used as a tool to guide a human expert who is trying to detect deception, not identify whether someone is lying or being truthful. In the future, the government should pilot test these types of deep learning models that detect contradictions for potential use as one tool to help guide security clearance background investigations.

Abbreviations

AWS	Amazon Web Services
CI	cognitive interviewing
FBI	Federal Bureau of Investigation
HITL	human-in-the-loop
LIWC	Linguistic Inquiry and Word Count
MCI	modified cognitive interviewing
ML	machine learning
NLP	natural language processing
SF-86	Standard Form-86
SNLI	Stanford Natural Language Inference
VTC	video teleconference

References

Alowibdi, Jalal S., Ugo A. Buy, Philip S. Yu, Sohaib Ghani, and Mohamed Mokbel, "Deception Detection in Twitter," *Social Network Analysis and Mining*, Vol. 5, 2015.

Amazon Web Services, "Amazon Transcribe," webpage, undated. As of November 21, 2021: https://aws.amazon.com/transcribe/

Arciuli, Joanne, David Mallard, and Gina Villar, "'Um, I Can Tell You're Lying': Linguistic Markers of Deception Versus Truth-Telling in Speech," *Applied Psycholinguistics*, Vol. 31, No. 3, July 2010, pp. 397–411.

Battestini, Agathe, Vidya Setlur, and Timothy Sohn, "A Large Scale Study of Text Messaging Use," *Proceedings of the 12th International Conference on Human Computer Interaction with Mobile Devices and Services*, September 2010, pp. 229–238.

Bond, Charles F., Jr., and Bella M. DePaulo, "Accuracy of Deception Judgments," *Personality and Social Psychology Review*, Vol. 10, No. 3, 2006, pp. 214–234.

Bond, Charles F., Jr., and Bella M. DePaulo, "Individual Differences in Judging Deception: Accuracy and Bias," *Psychological Bulletin*, Vol. 134, No. 4, July 2008, pp. 477–492.

Bower, Gordon H., "A Multicomponent Theory of the Memory Trace," in K. W. Spence and J. T. Spence, eds., *Psychology of Learning and Motivation: Advances in Research and Theory*, Vol. 1, New York: Academic Press, 1967.

Bowman, Samuel R., Gabor Angeli, Christopher Potts, and Christopher D. Manning, "A Large Annotated Corpus for Learning Natural Language Inference," *arXiv*, 1508.05326, August 21, 2015.

Capraro, Valerio, "Gender Differences in Lying in Sender-Receiver Games: A Meta-Analysis," *Judgment and Decision Making*, Vol. 13, No. 4, July 2018, pp. 345–355.

Clark, Herbert H., and Jean E. Fox Tree, "Using *Uh* and *Um* in Spontaneous Speaking," *Cognition*, Vol. 84, No. 1, May 2002, pp. 73–111.

Colomb, Cindy and Magali Ginet, "The Cognitive Interview for Use with Adults: An Empirical Test of an Alternative Mnemonic and of a Partial Protocol," *Applied Cognitive Psychology*, Vol. 26, No. 1, January–February 2012, pp. 35–47.

De Ruiter, Bob, and George Kachergis, "The Mafiascum Dataset: A Large Text Corpus for Deception Detection," *arXiv*, 1811.07851, last revised August 14, 2019.

DePaulo, Bella M., Deborah A. Kashy, Susan E. Kirkendol, Melissa M. Wyer, and Jennifer A. Epstein, "Lying in Everyday Life," *Journal of Personality and Social Psychology*, Vol. 70, No. 5, May 1996, pp. 979–995.

DePaulo, Bella M., James J. Lindsay, Brian E. Malone, Laura Muhlenbruck, Kelly Charlton, and Harris Cooper, "Cues to Deception," *Psychological Bulletin*, Vol. 129, No. 1, January 2003, pp. 74–118.

Dou, Jason, Michelle Liu, Haaris Muneer, and Adam Schlussel, "What Words Do We Use to Lie? Word Choice in Deceptive Messages," *arXiv*, 1710.00273, October 1, 2017.

Dreber, Anna, and Magnus Johannesson, "Gender Differences in Deception," *Economics Letters*, Vol. 99, No. 1, April 2008, pp. 197–199.

Enos, Frank, Stefan Benus, Robin L. Cautin, Martin Graciarena, Julia Hirschberg, and Elizabeth Shriberg, "Human Detection of Deceptive Speech," 2006.

Erat, Sanjiv, and Uri Gneezy, "White Lies," *Management Science*, Vol. 58, No. 4, 2011, pp. 723–733.

Geiselman, R. E, R. P. Fisher, I. Firstenberg, L. A. Hutton, S. J. Sullivan, I. V. Avetissain, and A. L. Prosk, "Enhancement of Eyewitness Memory—An Empirical Evaluation of the Cognitive Interview," *Journal of Police Science and Administration*, Vol. 12, No. 1, March 1984, pp. 74–80.

Hancock, Jeffrey T., Jeremy Birnholtz, Natalya Bazarova, Jamie Guillory, Josh Perlin, and Barrett Amos, "Butler Lies: Awareness, Deception, and Design," *Proceedings of the SIGCHI Conference on Human Factors in Computing Systems*, April 2009, pp. 517–526.

Hancock, Jeffrey T., Lauren E. Curry, Saurabh Goorha, and Michael Woodworth, "On Lying and Being Lied to: A Linguistic Analysis of Deception in Computer-Mediated Communication," *Discourse Processes*, Vol. 45, No. 1, 2008, pp. 1–23.

Hartwig, Maria, Pär Anders Granhag, and Leif A. Strömwall, "Guilty and Innocent Suspects' Strategies During Police Interrogations," *Psychology, Crime and Law*, Vol. 13, No. 2, 2007, pp. 213–227.

Hartwig, Maria, and Charles F. Bond Jr., "Why Do Lie-Catchers Fail? A Lens Model Meta-Analysis of Human Lie Judgments," *Psychological Bulletin*, Vol 137, No. 4, July 2011, pp. 643–659.

Hartwig, Maria, and Charles F. Bond, Jr., "Lie Detection from Multiple Cues: A Meta-Analysis," *Applied Cognitive Psychology*, Vol. 28, No. 5, 2014, pp. 661–676.

IBM, "Random Forest," webpage, December 7, 2020. As of August 2, 2022:
https://www.ibm.com/cloud/learn/random-forest

Johnson, Christian, and William Marcellino, *Bag-of-Words Algorithms Can Supplement Transformer Sequence Classification and Improve Model Interpretability*, Santa Monica, Calif.: RAND Corporation, WR-A1719-1, 2022. As of January 13, 2022:
https://www.rand.org/pubs/working_papers/WRA1719-1.html

Jung, Seeun and Radu Vranceanu, "Experimental Evidence on Gender Differences in Lying Behaviour," *Revue Économique*, Vol. 68, No. 5, 2017, pp. 859–873.

Khan, Wasiq, Keeley Crockett, James O'Shea, Abir Hussain, and Bilal M. Khan, "Deception in the Eyes of Deceiver: A Computer Vision and Machine Learning Based Automated Deception Detection," *Expert Systems with Applications*, Vol. 169, May 1, 2021, 114341.

Korte, Gregory, "'I Have Done Nothing Wrong'," *USA Today,* June 10, 2013.

Kowal, Sabine, Daniel C. O'Connell, Kathryn Forbush, Mark Higgins, Lindsay Clarke, and Karey D'Anna, "Interplay of Literacy and Orality in Inaugural Rhetoric," *Journal of Psycholinguistic Research*, Vol. 26, No. 1, 1997, pp. 1–31.

Lai, Vivian, and Chenhao Tan, "On Human Predictions with Explanations and Predictions of Machine Learning Models: A Case Study on Deception Detection," *Proceedings of the Conference on Fairness, Accountability, and Transparency*, January 2019, pp. 29–38.

Landström, Sara, Pär Anders Granhag, and Maria Hartwig, "Witnesses Appearing Live Versus on Video: Effects on Observers' Perception, Veracity Assessments and Memory," *Applied Cognitive Psychology*, Vol. 19, No. 7, November 2005, pp. 913–933.

Leongómez, Juan David, Viktoria R. Mileva, Anthony C. Little, and S. Craig Roberts, "Perceived Differences in Social Status Between Speaker and Listener Affect the Speaker's Vocal Characteristics," *PLoS One*, Vol. 12, No. 6, June 14, 2017, e0179407.

Levitan, Sarah Ita, Angel Maredia, and Julia Hirschberg, "Linguistic Cues to Deception and Perceived Deception in Interview Dialogues," *Proceedings of the 2018 Conference of the North America Chapter of the Association for Computational Linguistics: Human Language Technology*, Vol. 1, 2018, pp. 1941–1950.

Lipps, Oliver and Georg Lutz, "Gender of Interviewer Effects in a Multitopic Centralized CATI Panel Survey," *Methods, Data, Analyses*, Vol. 11, No. 1, 2017, pp. 67–86.

Marcellino, William, Christian Johnson, Marek N. Posard, and Todd C. Helmus, *Foreign Interference in the 2020 Election: Tools for Detecting Online Election Interference*, Santa Monica, Calif.: RAND Corporation, RR-A704-2, 2020. As of February 1, 2022: https://www.rand.org/pubs/research_reports/RRA704-2.html

Marcellino, William, Todd C. Helmus, Joshua Kerrigan, Hilary Reininger, Rouslan I. Karimov, and Rebecca Ann Lawrence, *Detecting Conspiracy Theories on Social Media: Improving Machine Learning to Detect and Understand Online Conspiracy Theories*, Santa Monica, Calif.: RAND Corporation, RR-A676-1, 2021. As of February 1, 2022: https://www.rand.org/pubs/research_reports/RRA676-1.html

Mann, Samantha, Aldert Vrij, and Ray Bull, "Suspects, Lies, and Videotape: An Analysis of Authentic High-Stake Liars," *Law and Human Behavior*, Vol. 26, No. 3, 2002, pp. 365–376.

Mendels, Gideon, Sarah Ita Levitan, Kai-Zhan Lee, and Julia Hirschberg, "Hybrid Acoustic-Lexical Deep Learning Approach for Deception Detection," *Proceedings of Interspeech 2017*, 2017, pp. 1472–1476.

Morgan, Charles A., Yaron G. Rabinowitz, Deborah Hilts, Craig E. Weller, and Vladimir Coric, "Efficacy of Modified Cognitive Interviewing, Compared to Human Judgments in Detecting Deception Related to Bio-Threat Activities," *Journal of Strategic Security*, Vol. 6, No. 3, Fall 2013, pp. 100–119.

Morgan, Charles A., III, Yaron Rabinowitz, Robert Leidy, and Vladimir Coric, "Efficacy of Combining Interview Techniques in Detecting Deception Related to Bio-Threat Issues," *Behavioral Sciences and the Law*, Vol. 32, No. 3, May–June 2014, pp. 269–285.

National Counterintelligence and Security Center, *Fiscal Year 2017 Annual Report on Security Clearance Determinations*, Washington, D.C.: Office of the Director of National Intelligence, 2017.

National Counterintelligence and Security Center, *Fiscal Year 2019 Annual Report on Security Clearance Determinations: Congressional Tasking*, Washington, D.C.: Office of the Director of National Intelligence, April 2020.

National Institute of Standards and Technology, *NIST/SEMATECH e-Handbook of Statistical Methods*, Gaithersburg, Md.: U.S. Department of Commerce, last updated April 2012.

National Institute of Standards and Technology, "1.3.5.16, Kolmogorov-Smirnov Goodness-of-Fit Test," *NIST/SEMATECH e-Handbook of Statistical Methods*, Gaithersburg, Md.: U.S. Department of Commerce, last updated April 2012.

Newman, Matthew L., James W. Pennebaker, Diane S. Berry, and Jane M. Richards, "Lying Words: Predicting Deception from Linguistic Styles," *Personality and Social Psychology Bulletin*, Vol. 29, No. 5, 2003, pp. 665–675.

Office of Management and Budget, "Questionnaire for National Security Positions," Standard Form 86, revised November 2016.

Palomäki, Jussi, Jeff Yan, David Modic, and Michael Laakasuo, "To Bluff Like a Man or Fold Like a Girl? Gender Biased Deceptive Behavior in Online Poker," *PLoS One*, Vol. 11, No. 7, 2016, e0157838.

Pérez-Rosas, Verónica, and Rada Mihalcea, "Experiments in Open Domain Deception Detection," *Proceedings of the 2015 Conference on Empirical Methods in Natural Language Processing*, September 2015, pp. 1120–1125.

Ringler, Hannah, Beata Beigman Klebanov, and David Kaufer, "Placing Writing Tasks in Local and Global Contexts: The Case of Argumentative Writing," *Journal of Writing Analytics*, Vol. 2, 2018, pp. 34–77.

Smith, Madeline E., Jeffrey T. Hancock, Lindsay Reynolds, and Jeremy Birnholtz, "Everyday Deception or a Few Prolific Liars? The Prevalence of Lies in Text Messaging," *Computers in Human Behavior*, Vol. 41, No. C, December 2014, pp. 220–227.

Sporer, Siegfried L., and Barbara Schwandt, "Moderators of Nonverbal Indicators of Deception: A Meta-Analytic Synthesis," *Psychology, Public Policy, and Law*, Vol. 13, No. 1, 2007, pp. 1–34.

Strömwall, Leif, and Pär Anders Granhag, "How to Detect Deception? Arresting the Beliefs of Police Officers, Prosecutors and Judges," *Psychology, Crime and Law*, Vol. 9, No. 1, 2003, pp. 19–36.

Strömwall, Leif, Maria Hartwig, and Pär Anders Granhag, "To Act Truthfully: Nonverbal Behaviour and Strategies During a Police Interrogation," *Psychology, Crime and Law*, Vol. 12, No. 2, 2006, pp. 207–219.

Tausczik, Yla R., and James W. Pennebaker, "The Psychological Meaning of Words: LIWC and Computerized Text Analysis Methods," *Journal of Language and Social Psychology*, Vol. 29, No. 1, 2010, pp. 24–54.

Toma, Catalina L., Jeffrey T. Hancock, and Nicole B. Ellison, "Separating Fact from Fiction: An Examination of Deceptive Self-Presentation in Online Dating Profiles," *Personality and Social Psychology Bulletin*, Vol. 34, No. 8, August 2008, pp. 1023–1036.

Tulving, Endel, and Donald M. Thomson, "Encoding Specificity and Retrieval Processes in Episodic Memory," *Psychological Review*, Vol. 80, No. 5, 1973, pp. 352–373.

Tulving, Endel, "Cue-Dependent Forgetting," *American Scientist*, Vol. 62, No. 1, January–February 1974, pp. 74–82.

Vaswani, Ashish, Noam Shazeer, Niki Parmar, Jakob Uszkoreit, Llion Jones, Aidan N. Gomez, Łukasz Kaiser, and Illia Polosukhin, "Attention Is All You Need," *Advances in Neural Information Processing Systems*, Vol. 30, 2017, pp. 5998–6008.

Villar, Gina, Joanne Arciuli, and David Mallard, "Use of 'Um' in the Deceptive Speech of a Convicted Murderer," *Applied Psycholinguistics*, Vol. 33, No. 1, January 2012, pp. 83–95.

Villar, Gina, and Paola Castillo, "The Presence of 'Um' as a Marker of Truthfulness in the Speech of TV Personalities," *Psychiatry, Psychology, and Law*, Vol. 24, No. 4, 2017, pp. 549–560.

Wetzel, Danielle, David Brown, Necia Werner, Suguru Ishizaki, and David Kaufer, "Computer-Assisted Rhetorical Analysis: Instructional Design and Formative Assessment Using DocuScope," *Journal of Writing Analytics*, Vol. 5, 2021, pp. 292–323.

Yancheva, Maria, and Frank Rudzicz, "Automatic Detection of Deception in Child-Produced Speech Using Syntactic Complexity Features," *Proceedings of the 51st Annual Meeting of the Association for Computational Linguistics*, Vol. 1: *Long Papers*, August 2013, pp. 944–953.

Zhou, Lina, Judee K. Burgoon, Jay F. Nunamaker Jr., and Doug Twitchell, "Automating Linguistics-Based Cues for Detecting Deception in Text-Based Asynchronous Computer-Mediated Communications," *Group Decision and Negotiation*, Vol. 13, No. 1, 2004, pp. 81–106.

Zuckerman, Miron, Bella M. DePaulo, and Robert Rosenthal, "Verbal and Nonverbal Communication of Deception," *Advances in Experimental Social Psychology*, Vol. 14, 1981, pp. 1–59.